Engineering Properties of Superconducting Materials

Engineering Properties of Superconducting Materials

Editor

Tim Coombs

MDPI • Basel • Beijing • Wuhan • Barcelona • Belgrade • Manchester • Tokyo • Cluj • Tianjin

Editor
Tim Coombs
Electrical Engineering Department,
Cambridge University
UK

Editorial Office
MDPI
St. Alban-Anlage 66
4052 Basel, Switzerland

This is a reprint of articles from the Special Issue published online in the open access journal *Materials* (ISSN 1996-1944) (available at: https://www.mdpi.com/journal/materials/special_issues/prop_supercond_mater).

For citation purposes, cite each article independently as indicated on the article page online and as indicated below:

LastName, A.A.; LastName, B.B.; LastName, C.C. Article Title. *Journal Name* **Year**, *Volume Number*, Page Range.

ISBN 978-3-0365-0372-1 (Hbk)
ISBN 978-3-0365-0373-8 (PDF)

© 2021 by the authors. Articles in this book are Open Access and distributed under the Creative Commons Attribution (CC BY) license, which allows users to download, copy and build upon published articles, as long as the author and publisher are properly credited, which ensures maximum dissemination and a wider impact of our publications.

The book as a whole is distributed by MDPI under the terms and conditions of the Creative Commons license CC BY-NC-ND.

Contents

About the Editor . . . . **vii**

Preface to "Engineering Properties of Superconducting Materials" . . . . **ix**

Tim Coombs
Engineering Properties of Superconducting Materials
Reprinted from: *Materials* **2020**, *13*, 4652, doi:10.3390/ma13204652 . . . . **1**

Lei Chen, Huiwen He, Guocheng Li, Hongkun Chen, Lei Wang, Xiaoyuan Chen, Xin Tian, Ying Xu, Li Ren and Yuejin Tang
Study of Resistive-Type Superconducting Fault Current Limiters for a Hybrid High Voltage Direct Current System
Reprinted from: *Materials* **2019**, *12*, 26, doi:10.3390/ma12010026 . . . . **3**

Kaixiang Liu, Lidong Dai, Heping Li, Haiying Hu, Linfei Yang, Chang Pu, Meiling Hong and Pengfei Liu
Phase Transition and Metallization of Orpiment by Raman Spectroscopy, Electrical Conductivity and Theoretical Calculation under High Pressure
Reprinted from: *Materials* **2019**, *12*, 784, doi:10.3390/ma12050784 . . . . **23**

Michael R. Koblischka Anjela Koblischka-Veneva Jörg Schmauch Masato Murakami
Microstructure and Flux Pinning of Reacted-and-Pressed, Polycrystalline $Ba_{0.6}K_{0.4}Fe_2As_2$ Powders
Reprinted from: *Materials* **2019**, *12*, 2173, doi:10.3390/ma12132173 . . . . **37**

Siyuan Liang, Li Ren, Tao Ma, Ying Xu, Yuejin Tang, Xiangyu Tan, Zheng Li, Guilun Chen, Sinian Yan, Zhiwei Cao, Jing Shi, Leishi Xiao and Meng Song
Study on Quenching Characteristics and Resistance Equivalent Estimation Method of Second-Generation High Temperature Superconducting Tape under Different Overcurrent
Reprinted from: *Materials* **2019**, *12*, 2374, doi:10.3390/ma12152374 . . . . **49**

Bright C. Robert, Muhammad U. Fareed and Harold S. Ruiz
How to Choose the Superconducting Material Law for the Modelling of 2G-HTS Coils
Reprinted from: *Materials* **2019**, *12*, 2679, doi:10.3390/ma12172679 . . . . **77**

Wenrong Li, Jie Sheng, Derong Qiu, Junbo Cheng, Haosheng Ye and Zhiyong Hong
Numerical Study on Transient State of Inductive Fault Current Limiter Based on Field-Circuit Coupling Method
Reprinted from: *Materials* **2019**, *12*, 2805, doi:10.3390/ma12172805 . . . . **97**

Ye Hong, Jun Zheng and Hengpei Liao
Modeling of High-T_c Superconducting Bulk using Different J_c–T Relationships over Dynamic Permanent Magnet Guideway
Reprinted from: *Materials* **2019**, *12*, 2915, doi:10.3390/ma12182915 . . . . **115**

About the Editor

Tim Coombs head the EPEC group in Cambridge and am a Fellow of both the Institute of Physics and the Institution of Engineering and Technology and a Senior Member of the IEEE. I have over 200 publications and a current H index of 30 (Scopus). I collaborate closely with industry on problems of applied superconductivity, both through academic research projects and, via my company Magnifye, through consultancy and product development. My research concentrates on applied superconductivity. I have made significant breakthroughs including superconducting bearings, maglev and fault current limiters for power distribution. This group works on many aspects of engineering applications of superconductivity and the fundamental physics which underpins them. I have published extensively on this in areas as diverse as AC losses, solutions to the critical state (we proposed and proved the H-formulation—now the industry standard), FCLS, magnets, motors, SMES, partial discharge, levitation, the effect of superconductors on the electricity network and, finally, the effect of flux pumping. I cover fundamental magnetism, electrical networks and the design and construction of magnets, motors and transformers. I have more than a dozen patents. I founded Magnifye Ltd. (www.magnifye.com) in 2007. Magnifye holds four major worldwide patents. I have made appearances on the BBC, Reuters and radio.

Preface to "Engineering Properties of Superconducting Materials"

Taking a technology from the laboratory to industry is a long and resource-consuming process. Discovered more than a century ago, the phenomenon of superconductivity is testament to this process. Despite the promise of this technology, currently, the only major use of superconductors outside the laboratory is in MRI machines. The advent of high-temperature superconductors in 1986 heralded a new dawn. Machines which do not require cooling with liquid helium are a very attractive target. A myriad range of different superconductors were rapidly discovered over the next decade. This process of discovery continues to this day with, most recently, a whole new class, the pnictides, being discovered in 2006. Many different usages have been identified, including in motors, generators, wind turbines, fault current limiters, and high-current low-loss cables. This Special Issue looks at some of the different factors which will help to realise these devices and thereby bring about a superconducting world.

Tim Coombs
Editor

Editorial

Engineering Properties of Superconducting Materials

Tim Coombs

Electrical Engineering Department, Cambridge University, Cambridge CB3 0FA, UK; tac1000@eng.cam.ac.uk

Received: 8 October 2020; Accepted: 13 October 2020; Published: 19 October 2020

Abstract: Taking a technology from the laboratory to industry is a long and resource-consuming process. Discovered more than a century ago, the phenomenon of superconductivity is testament to this process. Despite the promise of this technology, currently the only major use of superconductors outside the laboratory is in MRI machines. The advent of high-temperature superconductors in 1986 heralded a new dawn. Machines which do not require cooling with liquid helium are a very attractive target. A myriad range of different superconductors were rapidly discovered over the next decade. This process of discovery continues to this day with, most recently, a whole new class, the pnictides, being discovered in 2006. Many different usages have been identified, including in motors, generators, wind turbines, fault current limiters, and high-current low-loss cables. This Special Issue looks at some of the different factors which will help to realise these devices and thereby bring about a superconducting world

Keywords: HTS; bulk superconductors; coated conductors; mathematical modelling; H-formulation

---

The search for clean energy sources has been a fundamental key in materials research. The development of superconducting materials attracts significant scientific and technological resources towards achieving low costs, as well as suitable and profitable power generation, storage, distribution and transmission. In addition, superconducting electronics can provide devices and circuits with properties not obtainable by any other known technology; i.e., very low-loss, zero frequency-dispersion signal transmission lines, very high-Q-value resonators and filters, and quantum limited electromagnetic sensors.

All of these advances require high quality superconducting materials, and, in recent years, great strides have been made to improve the properties of existing materials, as well as the continuing discovery of new systems and materials, such as the pnictides.

In 1911, Heike Kamerlingh Onnes discovered superconductivity in mercury by cooling it down to a frosty 4.2 K (−268.95 °C). Since then, it has been the Holy Grail of material scientists to achieve this transition—from a normal to superconducting state—at room temperature (above 273.15 K or 0 °C). The hope of finding a room-temperature superconductor (RTS) arose after physicists discovered high-temperature superconductivity (HTS) in the 1980s and 1990s in a class of ceramic materials called cuprates [1]. They are characterised by the presence of interleaving copper-oxide layers. Their transition temperature—also known as critical temperature (Tc)—was significantly higher than those of conventional metallic superconductors discovered decades earlier. Even more recently MgB2 and a whole new family based on pnictides [2] were discovered.

There is a continuous drive towards higher and higher transition temperatures and to date, the highest superconducting Tc achieved, and confirmed, is 203 K, in 2015. From an engineering point of view, although higher transition temperatures are desirable, of greater interest is the development of the engineering properties of the materials.

The articles in this Special Issue reflect the broad nature of the subject and of the materials available to engineers. Superconductors are available as thin and thick films, single crystals, bulks, and tape. Of this large range, two of the most important to engineers are bulks and tapes.

Bulks have been available since the very early days of HTS [3]. In a high-quality bulk, the current can flow on the length scale of the superconductor and this property means they can operate as high-field 'permanent' magnets [4–6]. The field is only limited by the structural properties of the bulks and the source being used to magnetise them. Bulks are typically used in levitation, i.e., bearings or levitated trains. The best bulks have strong pinning properties as the stronger the pinning, the higher the field that they can maintain, and there is a continuous drive towards improved pinning properties.

Tapes, commonly known as coated conductors [7], are more versatile as they can be used to construct coils and hence act as magnets or to carry power at very high power densities. In addition, the property that once the critical current is exceeded they start to develop resistance can both be a blessing in a fault current limiter [8] (where the resistance is desirable and limits the current) and a curse in a magnet where the problem of quenching can cause irreparable damage [9]. Both of these issues are covered in this Special Issue.

The final topic covered in this Special Issue highlights the enormous effort which has been devoted to predicting the behaviour of superconductors by ever more sophisticated modelling techniques [10].

References

1. Bednorz, J.G.; Müller, K.A. Possible high T_c superconductivity in the Ba−La−Cu−O system. *Z. Phys. B Condens. Matter* **1986**, *64*, 189–193. [CrossRef]
2. Kamihara, Y.; Hiramatsu, H.; Hirano, M.; Kawamura, R.; Yanagi, H.; Kamiya, T.; Hosono, H. Iron-Based Layered Superconductor: LaOFeP. *J. Am. Chem. Soc.* **2006**, *128*, 10012–10013. [CrossRef] [PubMed]
3. Murakami, M. Novel application of high T_c bulk superconductors. *Appl. Supercond.* **1993**, *1*, 1157–1173. [CrossRef]
4. Tomita, M.; Murakami, M. High-temperature superconductor bulk magnets that can trap magnetic fields of over 17 tesla at 29 K. *Nature* **2003**, *421*, 517–520. [CrossRef] [PubMed]
5. Durrell, J.H.; Dennis, A.R.; Jaroszynski, J.; Ainslie, M.D.; Palmer, K.G.B.; Shi, Y.-H.; Campbell, A.M.; Hull, J.; Strasik, M.; Hellstrom, E.E. *Supercond. Sci. Technol.* **2014**, *27*, 082001.
6. Patel, A.; Baskys, A.; Mitchell-Williams, T.; McCaul, A.; Coniglio, W.; Hänisch, J.; Lao, M.; Glowacki, B.A. *Supercond. Sci. Technol.* **2018**, *31*, 09LT01.
7. Iijima, Y.; Matsumoto, K. High-temperature-superconductor coated conductors: Technical progress in Japan. *Supercond. Sci. Technol.* **2000**, *13*, 68–81. [CrossRef]
8. Zhang, X.; Ruiz, H.S.; Geng, J.; Shen, B.; Fu, L.; Zhang, H.; A Coombs, T. Power flow analysis and optimal locations of resistive type superconducting fault current limiters. *SpringerPlus* **2016**, *5*, 1972. [CrossRef] [PubMed]
9. Lebrun, P. Interim Summary Report on the Analysis of the 19 September 2008 Incident at the LHC. CERN EDMS document, 2008, 973073. Available online: https://www.slideshare.net/Intilla/interim-summary-report-on-the-analysis-of-the-19-september-2008-incident-at-the-lhc-cern-presentation (accessed on 25 September 2020).
10. Hong, Z.; Campbell, A.M.; Coombs, T.A. Computer Modelling of Magnetisation in High Temperature Bulk Superconductors. *IEEE Trans Appl. Supercond.* **2007**, *17*, 3761. [CrossRef]

Publisher's Note: MDPI stays neutral with regard to jurisdictional claims in published maps and institutional affiliations.

© 2020 by the author. Licensee MDPI, Basel, Switzerland. This article is an open access article distributed under the terms and conditions of the Creative Commons Attribution (CC BY) license (http://creativecommons.org/licenses/by/4.0/).

Article

Study of Resistive-Type Superconducting Fault Current Limiters for a Hybrid High Voltage Direct Current System

Lei Chen [1,*], Huiwen He [2], Guocheng Li [1], Hongkun Chen [1], Lei Wang [2], Xiaoyuan Chen [3], Xin Tian [4], Ying Xu [5], Li Ren [5] and Yuejin Tang [5]

1. School of Electrical Engineering and Automation, Wuhan University, Wuhan 430072, China; li_guo_cheng@163.com (G.L.); chkinsz@163.com (H.C.)
2. State Key Laboratory of Power Grid Environmental Protection, China Electric Power Research Institute, Wuhan 430074, China; husthhw@126.com (H.H.); wanglei8@epri.sgcc.com.cn (L.W.)
3. School of Engineering, Sichuan Normal University, Chengdu 610101, China; chenxy@sicnu.edu.cn
4. School of Electronic Information, Wuhan University, Wuhan 430072, China; xin.tian@whu.edu.cn
5. State Key Laboratory of Advanced Electromagnetic Engineering and Technology, Huazhong University of Science and Technology, Wuhan 430074, China; xuying@hust.edu.cn (Y.X.); renli@mail.hust.edu.cn (L.R.); tangyj@mail.hust.edu.cn (Y.T.)
* Correspondence: chen_lei@whu.edu.cn; Tel.: +86-135-1720-5365

Received: 24 November 2018; Accepted: 18 December 2018; Published: 21 December 2018

Abstract: In this paper, a hybrid high voltage direct current transmission system containing a line commutated converter and a voltage source converter is developed. To enhance the robustness of the hybrid transmission system against direct current short-circuit faults, resistive-type superconducting fault current limiters are applied, and the effectiveness of this approach is assessed. Related mathematical models are built, and the theoretical functions of the proposed approach are expounded. According to the transient simulations in MATLAB software, the results demonstrate that: (i) The superconducting fault current limiter at the voltage source converter station enables to very efficiently mitigate the fault transients, and owns an enhanced current-limiting ability for handling the short-line faults. (ii) The superconducting fault current limiter at the line commutated converter station is able to mildly limit the fault current and alleviate the voltage drop, and its working performance has a low sensitivity to the fault location. At the end of the study, a brief scheme design of the resistive-type superconducting fault current limiters is achieved. In conclusion, the application feasibility of the proposed approach is well confirmed.

Keywords: hybrid high voltage direct current transmission system; resistive-type superconducting fault current limiter; scheme design; short-circuit fault; Yttrium barium copper oxide materials; transient simulation

1. Introduction

In recent years, hybrid high voltage direct voltage (HVDC) technology has received continuously increasing attention, and it is known as an advanced option for long distance as well as large-scale power transmission [1,2]. In principle, a hybrid HVDC transmission system contains a line commutated converter (LCC) and a voltage source converter (VSC). The LCC serves as a rectifier station to save the capital cost, and the VSC acts as an inverter station to strengthen the operational flexibility. Due to integrating the merits of the LCC and VSC, the hybrid HVDC owns the following technical characteristics: (i) inexistence of commutation failure, (ii) enhanced competence to support weak/passive networks, (iii) flexible control of active and reactive power.

For promoting the development of hybrid HVDC technology, scholars have conducted some fundamental researches, which focus on the measure of alternating current (AC) system strength and the small-signal dynamics [3,4]. However, there are little studies on enhancing the robustness of a hybrid HVDC transmission system against DC short-circuit faults. In a sense, the hybrid HVDC may have a complex fault issue, since the DC fault currents of the LCC and VSC stations have essential differences with each other. For the DC fault current of the LCC station, it could be properly adjusted by the firing angle controller, and applying an additional current-limiting solution is able to bring a better fault suppression effect. Concerning the DC fault current of the VSC station, it rises very fast and cannot be removed even though the power electronic switches are blocked, while the anti-parallel diodes act as a freewheeling bridge circuit to feed the fault current [5]. Thus, it becomes an urgent and inevitable requirement to introduce an efficient current-limiting approach in the VSC station.

In this study, our research group suggests using superconducting fault current limiters (SFCLs) to solve the DC fault issue in the hybrid HVDC, and it is because SFCL is a very competitive current-limiting device with excellent performance superiorities, such as automatic trigger and rapid response [6–10]. Based on a comprehensive literature review, Table 1 lists a summary of the studies of SFCLs in different HVDC networks. Technically speaking, the current studies are mainly concerned about a pure LCC-HVDC or VSC-HVDC network.

In [11], a flux-coupling-type SFCL is applied to address the commutation failure in a pure LCC-HVDC grid. In light of different fault types and fault resistances, this SFCL's impacts on reducing the duration of the commutation failure and accelerating the fault recovery are confirmed. In [12,13], the performance behaviors of the resistive type SFCL on mitigating the commutation failure of a LCC-HVDC grid are studied. Different installation sites of the SFCL for the HVDC network are assessed, and a suitable optimal method of the SFCL resistance is investigated.

In [14–17], the SFCLs such as resistive-type, saturated iron-core-type, and hybrid-type are selected to inhibit the DC fault current of a pure VSC-HVDC network. A few helpful contributions regarding the parameter optimization and techno-economic evolution of the SFCLs for the VSC-HVDC network protection are obtained. In addition, some scholars preliminarily explore the influences of the resistive and inductive SFCLs on improving the operation reliability [18], and strengthening the fault ride-through (FRT) of wind plants connected to the VSC-HVDC network [19].

Table 1. Summary of the studies of superconducting fault current limiters (SFCLs) in different high voltage direct voltage (HVDC) networks. LCC: line commutated converter; VSC: voltage source converter.

Type of SFCL	Type of HVDC	Voltage Class	Research Object	Evaluation Method	Country, Report Year
Flux-coupling-type	LCC HVDC	230 kV	Commutation failure and fault recovery	MATLAB Simulation	China, 2015 [11]
Resistive type	LCC HVDC	180 kV [12] 500 kV [13]	Commutation failure and position analysis	PSCAD Simulation	Korea, 2016 [12] China, 2017 [13]
Hybrid-type	VSC HVDC	160 kV	Principle verification and scheme design	MATLAB Simulation	China, 2017 [14]
Resistive type	VSC HVDC	320 kV [15,19] 100 kV [16] 200 kV [18]	Techno-economic evolution and resistance varying behaviors	PSCAD [15,19] MATLAB [16,18]	France, 2017 [15] Korea, 2018 [16] China, 2017 [18] UAE, 2017 [19]
Inductive type	VSC HVDC	100 kV	Current limitation and recovery	MATLAB Simulation	Korea, 2018 [16]
Saturated iron-core-type	VSC HVDC	100 kV	Modeling, voltage analysis and energy dissipation	MATLAB Simulation	Korea, 2018 [16] China, 2018 [17]

To the best of our knowledge, there are no related reports about the systematic application of resistive SFCLs in a hybrid 500 kV HVDC network. When two resistive SFCLs are respectively installed at the LCC and VSC stations to withstand the DC fault, it is crucial to investigate how the two SFCLs can protect a hybrid HVDC network subject to the change of current-limiting parameters, fault resistances, and fault locations. In addition, it is critical to clarify the performance differences between the two SFCLs and lay a foundation for the scheme design of superconducting devices.

Aiming at the aforementioned tasks, this paper is devoted to studying and assessing the application feasibility of resistive SFCLs in a 500 kV hybrid HVDC network. The paper is arranged as follows. Section 2 states the analytical model of the hybrid HVDC including the SFCLs, and discusses the theoretical functions of the SFCLs to the DC fault behaviors. Section 3 conducts the simulation analyses and performance comparison, where different current-limiting parameters, fault severity levels and fault locations are taken into consideration. In Section 4, a brief scheme design of the SFCLs basing YBCO material is given. Section 5 recaps the main conclusions and suggests improvements in the future.

2. Theoretical Analysis

2.1. Analytical Model of the Hybrid HVDC Including the SFCLs

Figure 1 indicates the schematic connection of the hybrid HVDC system including two resistive SFCLs. The system is a 500 kV bipolar hybrid LCC-VSC HVDC link (only positive pole is denoted here), and the two SFCLs are installed at the LCC station (rectifier side) and the VSC station (inverter side), respectively. For the analytical model of the hybrid HVDC system, this study mainly considers the following factors: (i) The two AC grids are represented by equivalent AC voltage sources with series impedances [20]. (ii) The DC transmission line is represented by an equivalent "resistance-inductance (R-L)" model. (iii) The LCC station adopts a constant DC current control to generate the firing angle, and the VSC station uses a direct current control mode [21]. Detailed modeling information as well as mathematical equations can be found in Appendices A.1–A.3.

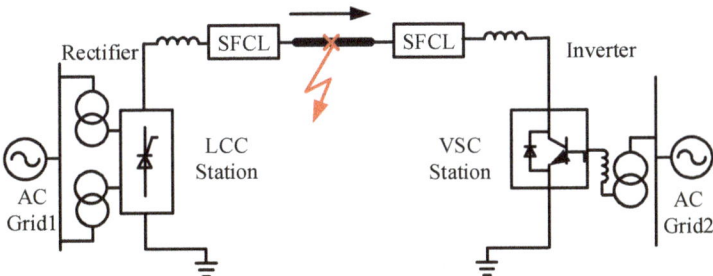

Figure 1. Allocation of the resistive SFCLs in a hybrid HVDC system with the LCC and VSC station.

Based on the second generation Yttrium barium copper oxide (YBCO) material, Figure 2 describes the equivalent modeling of a resistive SFCL at different time-scales [16,22,23], and the change rule of the SFCL resistance is written as:

$$R(t) = \begin{cases} 0 & (t < t_0) \\ R_{SC}[1 - \exp(-\frac{t-t_0}{\tau})]^{\frac{1}{2}} & (t_0 \leq t < t_1) \\ a_1(t - t_1) + b_1 & (t_1 \leq t < t_2) \\ a_2(t - t_2) + b_2 & (t_2 \leq t < t_3) \end{cases} \quad (1)$$

where t is the time constant; R_{SC} is denoted as the normal-state resistance of the SFCL. From Equation (1), the SFCL's equivalent model is explained as: (i) t_0 is the quench-starting time; t_1 is the first-stage recovery-starting time; t_2 is the secondary-stage recovery-starting time; t_3 is the completed recovery time. (ii) a_1, a_2, b_1, and b_2 are expressed as the model coefficients, respectively.

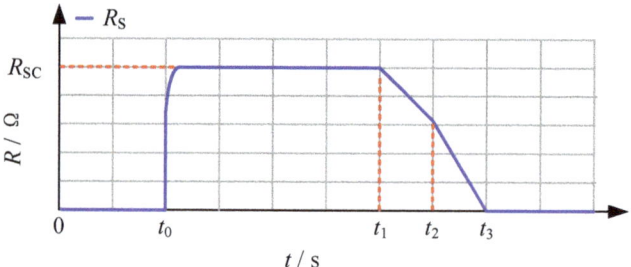

Figure 2. The simplified mathematical model of the resistive SFCL.

From the literature, the modeling of the two-stage recovery time is mainly based on the experimental studies for superconducting elements, and the temperature effect could be the potential reason. When the SFCL starts to recover to the superconducting state, the accumulated joule heat under the current-limiting operation leads to the temperature rising and make the SFCL resistance have a constrained variation trend, and it is defined as the first-stage recovery process. After the heat is dissipated, the lowering of temperature will make the SFCL resistance have a faster drop to zero, and it is defined as the two-stage recovery process. As for a precise resistive SFCL model on account of the "power-law" equation, this equivalent SFCL model has multiple operational segments, and it is still valid to reflect the transient properties of the resistive SFCL.

2.2. Impacts of the SFCLs on the DC Fault Currents

In this section, the impacts of the SFCLs on the DC fault currents of the LCC and VSC stations are discussed. As shown in Figure 3, it indicates the equivalent circuit of the DC-link. U_{dcr}, I_{dcr} and U_{dci}, I_{dci} are represented as the DC voltage and current of the LCC and VSC stations, respectively; X_{smr} and X_{smi} are marked as the smoothing reactors installed at the LCC and VSC stations, respectively; C_{VSC} is the DC capacitor of the VSC station.

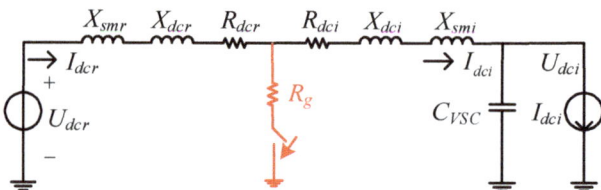

Figure 3. The equivalent circuit of the DC-link.

It is assumed that the fault resistance is R_g and the residual voltage at the fault location is U_g. Herein, R_{dcr}, X_{dcr}, and R_{dci}, X_{dci} are represented as the resistance and reactance of the DC line of the LCC and VSC stations, respectively. As the SFCL resistance R_{SFCLr} is connected in series with the LCC station, the dynamics of the DC fault current I_{dcr-f} can be expressed as:

$$(L_{smr} + L_{dcr})\dot{I}_{dcr-f} = U_{dcr} - U_g - I_{dcr-f}(R_{SFCLr} + R_{dcr}) \tag{2}$$

If the DC voltage U_{dcr} is represented by the function of firing angle, AC voltage and transformer leakage reactance, Equation (2) will be rewritten as:

$$\begin{cases} (L_{smr} + L_{dcr})\dot{I}_{dcr-f} = U_{dcr} - U_g - I_{dcr-f}(R_{SFCLr} + R_{dcr}) \\ U_{dcr} = 2(\frac{3\sqrt{2}V_{LCCm}}{\pi T}\cos\alpha - I_{dcr-f}\frac{3X_{T1}}{\pi}) \end{cases}$$
$$\Rightarrow (L_{smr} + L_{dcr})\dot{I}_{dcr-f} = \frac{6\sqrt{2}U_{LCCm}}{\pi T}\cos\alpha - U_g$$
$$-I_{dcr-f}(R_{SFCLr} + R_{dcr} + \frac{6X_{T1}}{\pi})$$
(3)

where U_{LCCm} is the root-mean-square (RMS) AC voltage over the LCC station; T is the transformer turn-ratio. As compared to the case of without SFCL, introducing R_{SFCLr} is able to increase the resistance of the DC circuit, and it is helpful to reduce the peak value of the fault current. Considering the function of the firing angle controller, the working status of the LCC station will be changed from the rectifying mode to the inverting mode. Thus, the DC fault current will be enforcedly down to zero.

Note that, the residual voltage U_g in Equation (2) can be calculated by:

$$U_g = R_g(I_{dcr-f} + I_{dci-f})$$
(4)

where I_{dci-f} is the DC fault current of the VSC station. It can be inferred that, when the ground resistance R_g is not equal to zero and has a relatively large resistance value, the DC fault current of the LCC station might be potentially affected by that of the VSC station.

By referring to [24–26], the fault process of a VSC-HVDC station has three stages, which are DC-link capacitor discharging (stage 1), diodes freewheeling (stage 2) and grid-side current feeding (stage 3), respectively. Before the DC voltage drops to zero, all the free-wheel diodes are blocked due to the reverse voltage, and thus the DC link will be insulated from the AC grid 2. In a sense, stage 1 (capacitor discharging) is the key stage for the SFCL to suppress the DC fault current and mitigate the DC voltage decline.

Herein, stage 1 conducts the system response before the dc voltage drops to zero, and Figure 4 shows the fault analysis diagram. The circuit equation is modeled as:

$$\begin{cases} (L_{smr} + L_{dcr})\dot{I}_{dci-f} = U_{dci} - U_g - I_{dci-f}(R_{SFCLi} + R_{dci}) \\ I_{dci-f} = -I_{Cap} = -C_{VSC}\dot{U}_{dci} \end{cases}$$
$$\Rightarrow -(L_{smr} + L_{dcr})C_{VSC}\ddot{U}_{dci} = U_{dci} - U_g + C_{VSC}\dot{U}_{dci}(R_{SFCLi} + R_{dci})$$
$$\Rightarrow (L_{smr} + L_{dcr})C_{VSC}\ddot{U}_{dci} + C_{VSC}\dot{U}_{dci}(R_{SFCLi} + R_{dci}) + U_{dci} - U_g = 0$$
(5)

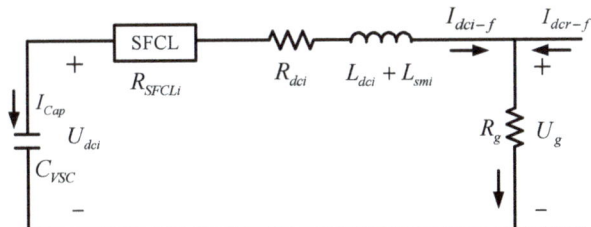

Figure 4. Fault analysis of the VSC-station with the SFCL (capacitor discharging stage).

By substituting Equation (4) into Equation (5), the equation will be rewritten as:

$$(L_{smr} + L_{dcr})C_{VSC}\ddot{U}_{dci} + C_{VSC}\dot{U}_{dci}(R_{SFCLi} + R_{dci} + R_g)$$
$$+ U_{dci} - R_g I_{dcr-f} = 0$$
(6)

Regarding the solution method of Equation (6), details are analyzed in Appendix ??. In theory, introducing R_{SFCLi} will closely affect the VSC-HVDC link's electrical properties, and it means that the current-limiting resistance R_{SFCLi} can not only reduce the fault current level in the DC line, but also change the oscillation characteristic of the DC capacitor voltage. When increasing R_{SFCLi} leads to an over-damped state, the capacitor voltage U_{dci} will not decline to zero, and the subsequent two stages may not happen [27,28]. Owing to that all-diodes-conducting phenomenon is avoided, the SFCL's contributions in reducing the currents in the AC side and the converter may become more obvious.

According to the above theoretical analysis, the flowchart of the integrated process of the proposed approach can be shown in Figure 5.

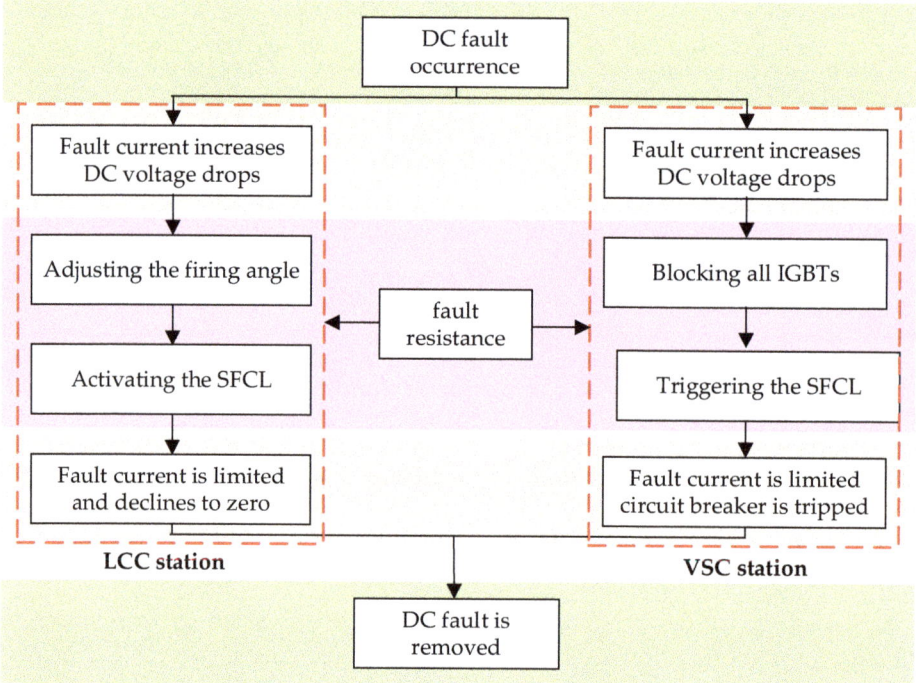

Figure 5. Flowchart of the integrated process of the proposed approach. IGBT: Insulated Gate Bipolar Translator.

3. Simulation Study

To evaluate the effectiveness of the SFCLs in the hybrid HVDC system, a detailed simulation model is built in the MATLAB software (R2017b, MathWorks, Natick, MA, USA), and the electromagnetic transient (EMT) type simulations are done in a 64-b personal computer with Intel i7-7700 QuadCore 2.8-GHz processor and 8-GB RAM (DELL, Round Rock, TX, USA). The EMT simulations use the discrete solver, and the simulation time step is set as 5×10^{-5} s.

The main parameters are summarized in Table 2, and the modeling information is depicted as: (i) The AC grid model is simulated by an AC voltage source in series with the equivalent resistance and inductance. (ii) The resistive SFCL model is based on the controlled voltage source [29]. (iii) The VSC and LCC adopt detailed models (detailed representation of power electronic converters), and the models are able to precisely show the dynamic performance over relatively short periods of times.

During the simulations, different SFCL resistances are taken into consideration [30,31], so as to validate how the change of the SFCL resistance affects the fault characteristics of the hybrid HVDC system. The estimated recovery time of the SFCL is about 4 s. For the resistance of R_{SFCL} = 30 Ω, the coefficients of a_1, a_2, b_1, b_2 are set as a_1 = 9.52, a_2 = 15.87, b_1 = 30, b_2 = 19, respectively.

Table 2. Main parameters of the simulation model.

Superconducting Fault Current Limiters	
Superconducting coil R_{sc} at the LCC/VSC	20 Ω–100 Ω/10 Ω–50 Ω
LCC Station	
Rated voltage/frequency	380 kV/50 Hz
Short-circuit ratio	3.076
DC current controller	K_{pIdc} = 1, K_{iIdc} = 90
DC Link	
Rated voltage/current	500kV/2 kA
Length of DC transmission line	500 km
Smoothing reactor of LCC/VSC	0.3 H/0.01 H
VSC Station	
Rated voltage/frequency	220 kV/50 Hz
Short-circuit ratio	3.34
AC current controller (K_{pVac}, K_{iVac})	K_{pVac} = 0.6, K_{iVac} = 10
DC voltage controller (K_{pVdc}, K_{iVdc})	K_{pVac} = 8, K_{iVac} = 20

3.1. Changing the SFCL Resistance in the LCC Station

The simulation conditions of the DC short-circuit fault are defined as: (i) The fault occurs in the middle of the DC line at t_0 = 3 s. (ii) The fault resistance and duration are 1 Ω and 100 ms. (iii) The SFCL at the LCC station (R_{SFCLr}) changes from 20 Ω to 100 Ω, and the SFCL at the VSC station (R_{SFCLi}) has the constant of 30 Ω. As shown in Figures 6 and 7, they indicate the transient behaviors of the hybrid HVDC system subject to the change of R_{SFCLr}.

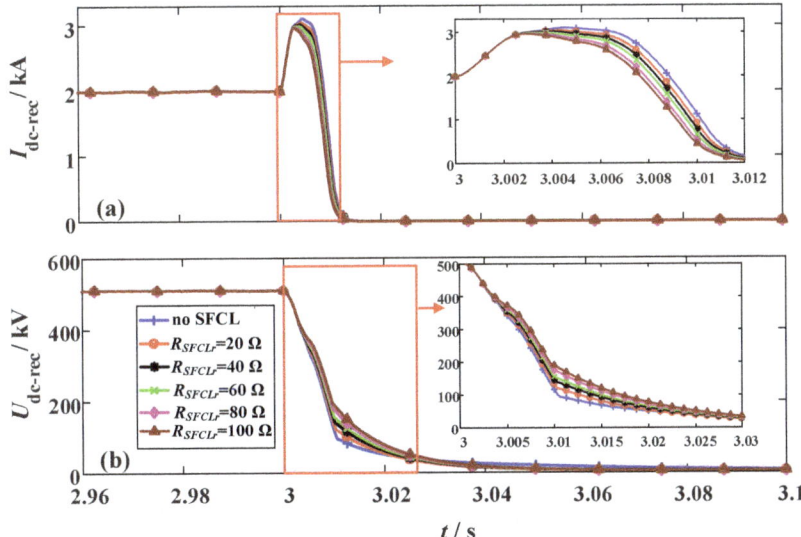

Figure 6. Behaviors of the LCC station considering the change of the SFCL resistance R_{SFCLr}. (**a**) DC current and (**b**) DC voltage.

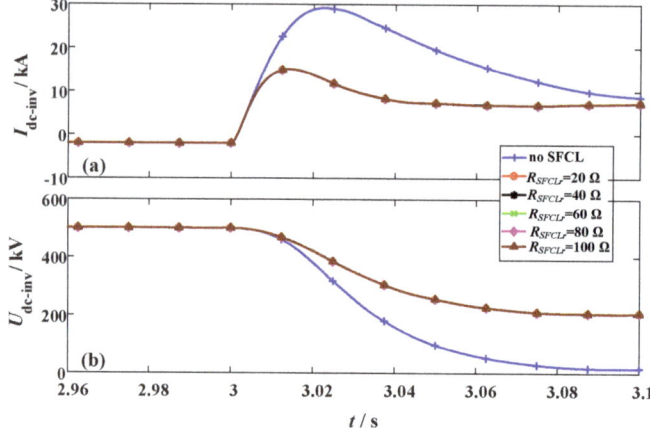

Figure 7. Behaviors of the VSC station considering the change of the SFCL resistance R_{SFCLr}. (**a**) DC current and (**b**) DC voltage.

In the LCC station, the peak value of the DC fault current is about 1.5 times of the rated level, and the reduction of the DC fault current is mainly conducted by adjusting the firing angle. Herein, the LCC station will switch to the inverting mode to make the DC fault current decline to zero. In the case of with the SFCL, it can mildly limit the fault current and alleviate the voltage drop. During the process of the fault feeding, the firing angle controller and the SFCL will serve as the primary and secondary factors to combinedly affect the fault transients.

It is observed that augmenting R_{SFCLr} has almost no effect on the VSC station, where the activation of R_{SFCLi} will undertake the crucial roles of current-limitation and voltage compensation. For the VSC station, installing the SFCL (R_{SFCLi} = 30 Ω) is able to limit the DC fault current from 29.2 kA to 15.1 kA, and improve the DC voltage from 16 kV to 197 kV.

Figure 8 shows the energy dissipation of the two SFCLs, where the calculation time is from the fault occurring to the fault being removed (the duration is 100 ms). When R_{SFCLr} is designed as 20 Ω, 40 Ω, 60 Ω, 80 Ω, and 100 Ω, respectively, its dissipated energy at the LCC station will be 0.27 MJ, 0.44 MJ, 0.56 MJ, 0.73 MJ, and 0.85 MJ, respectively. A rising trend is obviously found, but the caused energy dissipation effect is still limitable. In the VSC station, its relevant SFCL has a steady and efficient energy dissipation with the level of 119.9 MJ.

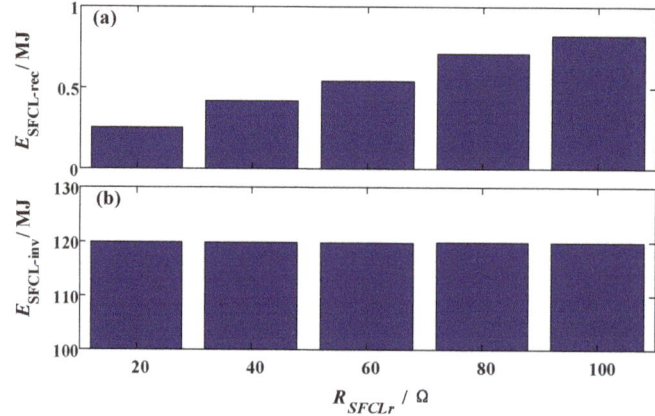

Figure 8. Energy dissipation of the two SFCLs subject to the change of the SFCL resistance R_{SFCLr}.

3.2. Changing the SFCL Resistance in the VSC Station

In this subsection, the original fault parameters are unchanged, and it is designed that R_{SFCLr} owns a constant of 20 Ω and R_{SFCLi} varies from 10 Ω to 50 Ω.

Figures 9 and 10 show the characteristics of the hybrid HVDC system subject to the change of R_{SFCLi}. According to the results, changing R_{SFCLi} will obviously influence the transient fluctuations in the VSC station, but have a negligible effect on the DC current and voltage of the LCC station. In addition, it is found that a moderate increase of the SFCL resistance R_{SFCLi} can bring better contributions. Nevertheless, it is not recommended to excessively increasing the resistance R_{SFCLi}, since the current-limiting ratio of the SFCL seems to achieve the saturated level. When R_{SFCLi} increases from 10 Ω to 20 Ω, the current-limiting ratio has an expected improvement of 14.3%, but when R_{SFCLi} rises from 40 Ω to 50 Ω, the obtained improvement is just 4.8%. As shown in Table 3, it lists a detailed performance comparison.

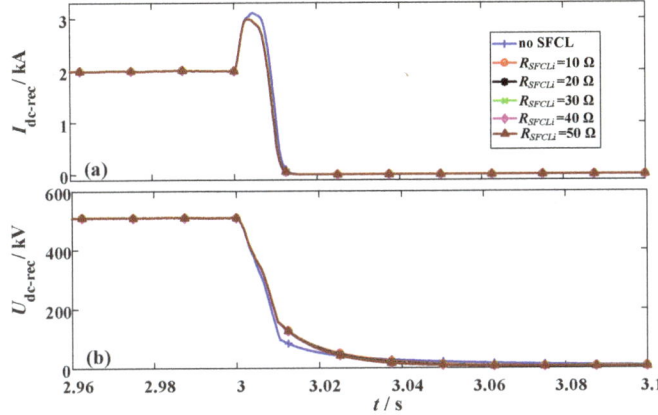

Figure 9. Behaviors of the LCC station considering the change of the SFCL resistance R_{SFCLi}. (**a**) DC current and (**b**) DC voltage.

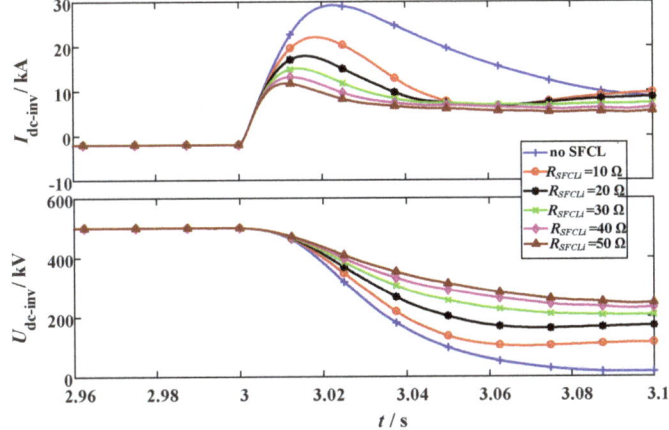

Figure 10. Behaviors of the VSC station considering the change of the SFCL resistance R_{SFCLi}. (**a**) DC current and (**b**) DC voltage.

Table 3. Performance of the SFCL at the VSC station under different parameters.

Items	Effects of the SFCL on the VSC Station	
	DC Fault Current/Current-Limiting Ratio	DC Voltage/Calculated Drop Rate
R_{SFCLi} = 10 Ω	22.1 kA/24.7%	101.5 kV/79.7%
R_{SFCLi} = 20 Ω	17.8 kA/39%	160.1 kV/67.9%
R_{SFCLi} = 30 Ω	15.1 kA/48.5%	196.9 kV/60.6%
R_{SFCLi} = 40 Ω	13.2 kA/55.1%	223.8 kV/55.2%
R_{SFCLi} = 50 Ω	11.7 kA/59.9%	241.3 kV/51.7%

Figure 11 shows the energy dissipation of the two SFCLs. Concerning that R_{SFCLi} is set as 10 Ω, 20 Ω, 30 Ω, 40 Ω, and 50 Ω, respectively, the dissipated energy of the SFCL at the VSC station will be 60.5 MJ, 96.1 MJ, 119.9 MJ, 128.0 MJ, and 129.3 MJ, respectively. Regarding the LCC station, its relevant SFCL has a steady energy dissipation with the level of 0.27 MJ.

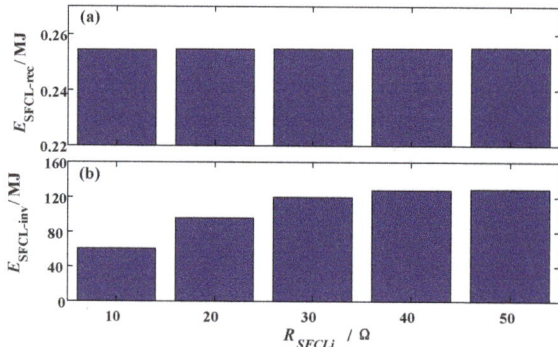

Figure 11. Energy dissipation of the two SFCLs subject to the change of the SFCL resistance R_{SFCLi}.

3.3. Changing the Fault Resistance of the Hybrid HVDC

As the fault resistance is a critical factor to evaluate the fault severity level and the behavioral interaction between the LCC and VSC stations, different fault resistances are simulated. The parameters of R_{SFCLr} = 20 Ω and R_{SFCLi} = 30 Ω are adopted, and the settings of the fault time and fault location are unchanged. The simulation waveforms are shown in Figures 12–14.

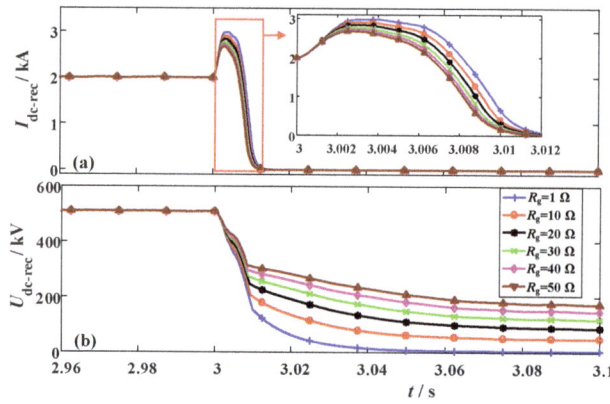

Figure 12. Properties of the LCC station considering the change of the fault resistance R_g. (a) DC current and (b) DC voltage.

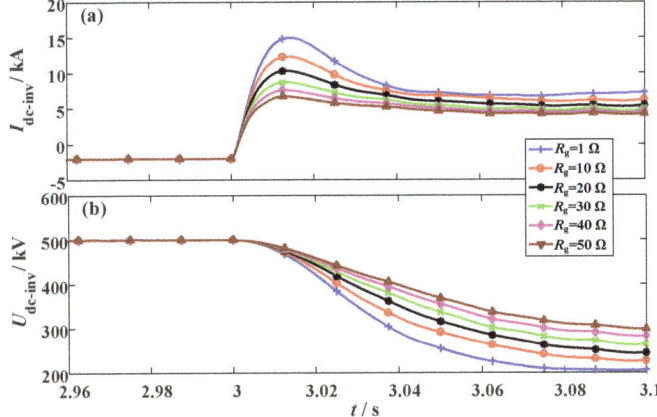

Figure 13. Properties of the VSC station considering the change of the fault resistance R_g. (**a**) DC current and (**b**) DC voltage.

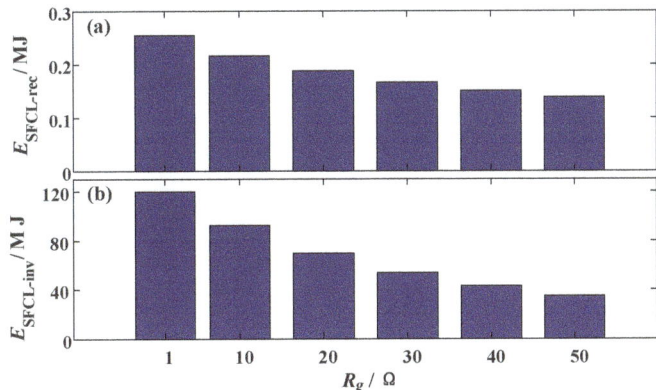

Figure 14. Energy dissipation of the two SFCLs subject to the change of the fault resistance R_g.

From Figure 12, the main contribution of augmenting the fault resistance for the LCC station is to mitigate the DC voltage drop. The DC fault current still decreases to zero, but the DC voltage can be properly kept owing to the voltage support over the fault resistance. In the case of that the fault resistance is set as 10 Ω, 20 Ω, 30 Ω, 40 Ω, and 50 Ω, respectively, the DC voltage will reach to 50.8 kV, 88.6 kV, 119.2 kV, 143.1 kV, and 164.7 kV, respectively.

From Figure 13, the DC current and voltage of the VSC station will be both affected by the fault resistance, and a detailed performance comparison is given in Table 4.

Table 4. Influence of the fault resistance on the VSC station.

Items	DC Fault Current		DC Voltage	
	No SFCL	With SFCL/Current-Limiting Ratio	No SFCL	With SFCL
R_g = 10 Ω	21.2 kA	12.4 kA/41.7%	99.1 kV	222.2 kV
R_g = 20 Ω	16.1 kA	10.3 kA/36.1%	159.1 kV	239.8 kV
R_g = 30 Ω	12.8 kA	8.76 kA/31.7%	196.9 kV	252.1 kV
R_g = 40 Ω	10.6 kA	7.62 kA/28.3 %	221.5 kV	261.8 kV
R_g = 50 Ω	9.03 kA	6.74 kA/25.4%	239.9 kV	271.9 kV

Owing to the increase of the fault resistance, the dissipated energies in the two SFCLs are both reduced. Especially for the SFCL at the VSC station, an evident downswing is observed. For that R_g is set as 10 Ω, 20 Ω, 30 Ω, 40 Ω, and 50 Ω, respectively, the dissipated energy of the SFCL at the VSC station will be 92.3 MJ, 69.7 MJ, 53.6 MJ, 42.9 MJ, and 34.9 MJ, respectively.

3.4. Changing the Fault Location of the Hybrid HVDC

To analyze how the fault location could affect the performance of the SFCLs, different fault sites are simulated. The fault location ratio is used to describe the relative position of the fault site in the whole DC line. When the fault location ratio increases, it means the fault site is farther away from the LCC station and closer to the VSC station. The two SFCLs still adopt R_{SFCLr} = 20 Ω and R_{SFCLi} = 30 Ω; the fault time and fault resistance are set as t_0 = 3 s and R_g = 1 Ω, respectively. The simulation waveforms are shown in Figures 15–17.

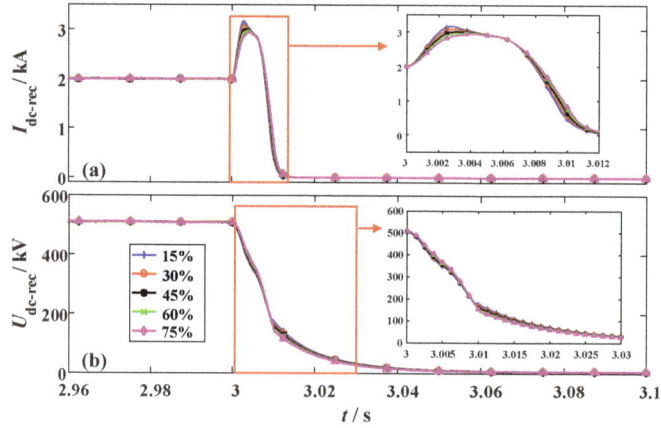

Figure 15. Simulation waveforms of the LCC station considering the change of the fault location. (**a**) DC current and (**b**) DC voltage.

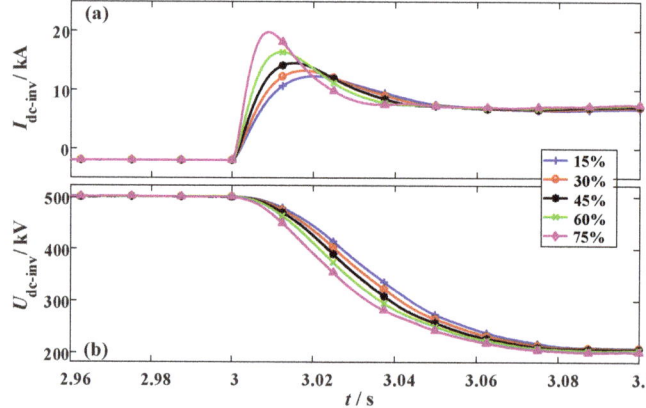

Figure 16. Simulation waveforms of the VSC station considering the change of the fault location. (**a**) DC current and (**b**) DC voltage.

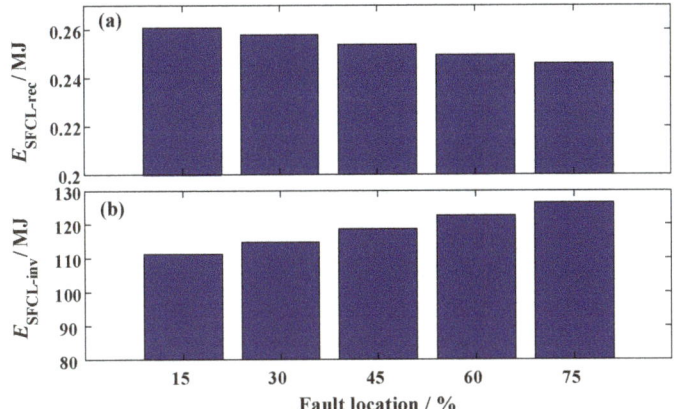

Figure 17. Energy dissipation of the two SFCLs subject to the change of the fault location.

When the fault location ratio changes from 15% to 75%, the peak value of the DC fault current in the LCC station just changes from 3.11 kA to 2.97 kA. Considering that the expected reduction is just 0.14 kA, the SFCL at the LCC station has a low sensitivity to the fault location. In comparison, the DC fault current of the VSC station is more sensitive to the fault site. When the fault location ratio is 15%, 30%, 45%, 60%, and 75%, respectively, the corresponding current-limiting ratio will be about 44.9%, 46.5%, 47.9%, 49.3%, and 50.4%, respectively. It is proven that the SFCL at the VSC station has an enhanced current-limiting ability for handling the short-line faults.

Based on Figure 17, Table 5 shows the simulation data of the two SFCLs' energy dissipation, whose changing trends are opposite with each other.

Table 5. Influence of the fault location on the energy dissipation of the two SFCLs.

Fault Location Ratio	Energy Dissipation	
	SFCL at the LCC Station	SFCL at the VSC Station
15%	0.265 MJ	111.2 MJ
30%	0.258 MJ	114.8 MJ
45%	0.252 MJ	118.7 MJ
60%	0.247 MJ	122.7 MJ
75%	0.243 MJ	126.5 MJ

4. Scheme Design

In this section, the SFCL scheme design is conducted. Firstly, the candidates for the structure of the resistive SFCL used in the HVDC networks are discussed. Figure 18a shows a general structure, which represents a pure resistive SFCL without an external resistor in parallel. Some scholars have applied this structure in [15,18], where the scholars consider the coordination of a high-speed direct-current circuit breaker (DCCB) and the SFCL. As the DCCB cuts off the DC fault current within 2–5 ms, the current-limiting time of the resistive SFCL can be controlled as 20 ms–50 ms. In light of a relatively short current-limiting time, the quench heat dissipation could be acceptable to a certain extent.

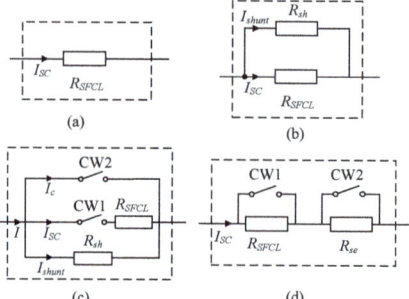

Figure 18. Candidates for the structure of the resistive SFCL used in the HVDC networks. (**a**) Pure resistive SFCL; (**b**) Resistive SFCL with external resistor in parallel; (**c**) Resistive SFCL with switches and external resistor in parallel; (**d**) Resistive SFCL with external resistor in series.

Figure 18b,c show two possible structures for the resistive SFCL with an external resistor in parallel [13,16,32]. In a sense, the scholars adopt a conservative and safe method, and the objective of introducing the external resistor is to avoid that the recovery process of the SFCL is too long. In addition, Figure 18d shows the structure of the resistive SFCL with an external resistor in series. The rating of the external resistor is the same as that of the resistive SFCL. When CW1 is closed and CW2 is opened, the external resistor will replace the SFCL to mitigate the fault transients. Hence, the current-limiting time of the resistive SFCL can be flexibly adjusted to ensure the safety and reliability of superconducting materials.

In this study, our research group prefers to use the general structure in Figure 18a. In case of this structure does not fully meet the requirement that the recovery time of the SFCL is about 4 s, the structure in Figure 18d can be regarded as an alternative solution. It should be noted that, the alternative structure may have the same current-limiting resistance as the preferred structure, and it does not affect the above simulation results of the DC current and voltage. In the following, the parameter selection is discussed.

For the LCC station, this study suggests installing the resistive SFCL with a lower quench resistance (no more than 20 Ω). On the one hand, it may cooperate with the firing angle controller to combinedly handle the DC fault issue. On the other hand, it may assist the SFCL at the VSC station to more powerfully handle the AC fault when the fault location is near the AC grid 1.

For the VSC station, this study recommends applying the resistive SFCL with R_{SFCLi} = 30 Ω, which is sufficient to alleviate the DC voltage-current fluctuations and dissipate the active power. As shown in Figure 19, the power response of the AC systems is demonstrated, and here the fault resistance is R_g = 1 Ω; the fault location is the middle of the DC transmission line.

Note that, it is not suggested to augment the SFCL resistance in excess. There might be a critical resistance value to depict the tradeoff among the SFCL cost, the fault current reduction and the inhibition of the voltage fluctuation [33]. Since detailed optimization and calculation are out of the scope of this paper, and will be presented in another report, a reasonable choice of R_{SFCLi} = 30 Ω is adopted to implement the SFCL's scheme design.

On basis of [34,35], a non-inductive unit coil for the SFCL is designed, and the coil parameters are listed in Table 6. To construct the SFCL at the VSC station, the normal current in the SFCL is 2 kA, and thus 15 pieces of coils connected in parallel are served as a coil group, which can meet the requirements of current capacity and safety margin. Further, 160 coil-groups connected in series is to obtain the quench resistance of 30 Ω.

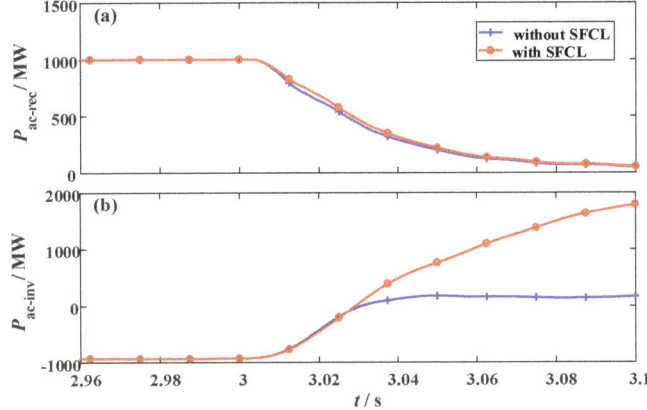

Figure 19. Power response of the AC systems under the fault. (**a**) AC grid 1 and (**b**) AC grid 2.

Table 6. Parameters of a non-inductive coil unit.

Parameter	Value
Length of YBCO tape (m)	54
Inner diameter (mm)	200
Outer diameter (mm)	870
Interturn gap (mm)	5
Resistance (Ω)	1.28 Ω @ 100 K
Resistance (Ω)	2.98 Ω @ 300 K
N Value (μV/cm)	38.6
Rated voltage (kV)	3
Break-down voltage (kV)	15
Rated current (A)	200
Peak current (A, with a duration of 100 ms)	900
Safety temperature limit (K)	300

5. Conclusions

In this paper, the application feasibility of resistive-type superconducting fault current limiters in a hybrid high voltage direct current transmission system is verified. The main conclusions are as follows:

1. The superconducting fault current limiter at the voltage source converter station enables to very efficiently mitigate the fault transients, and owns an enhanced current-limiting ability for handling the short-line faults. A moderate increase of the current-limiting resistance can bring better contributions, but an excessive increase may make the current-limiting ratio come up to the saturated level.
2. The superconducting fault current limiter at the line commutated converter station is able to mildly limit the fault current and alleviate the voltage drop, and its working performance has a low sensitivity to the fault location. As for the primary and secondary factors, the firing angle controller and the superconducting fault current limiter will combinedly handle the fault transients.

Concerning our future tasks, the optimization design, and economic evaluation of the superconducting fault current limiters will be done. Besides, the effects of the superconducting fault current limiters on the integration of large-scale renewable power sources into the hybrid system will be explored. These mentioned studies will be presented in the follow-up reports.

Author Contributions: L.C. helped design the study and write the manuscript; H.H. helped analyze the data; G.L. helped build the simulation model; H.C. helped conduct the study; L.W. helped analyze the HVDC model; X.C. helped conduct the device scheme; X.T. helped draw the figure; Y.X. helped conduct the simulation analysis; L.R. helped conduct the superconductor parameters; Y.T. helped edit the manuscript.

Funding: This research was funded by the Project of the SGCC under grant number GYW17201700027, the National Natural Science Foundation of China under grant number 51877154, 51507117, the Fundamental Research Funds for the Central Universities under grant number 2042018kf0214, the State Key Laboratory of Advanced Electromagnetic Engineering and Technology under grant number AEET 2018KF004.

Acknowledgments: The authors gratefully acknowledge financial support from the Project of the SGCC (GYW17201700027), the National Natural Science Foundation of China (51877154, 51507117), the Fundamental Research Funds for the Central Universities (2042018kf0214), the State Key Laboratory of Advanced Electromagnetic Engineering and Technology (AEET 2018KF004).

Conflicts of Interest: The authors declare no conflict of interest.

Nomenclature

LCC	Line commutated converter
VSC	Voltage source converter
RMS	Root mean square
FRT	Fault ride-through
HVDC	High voltage direct current
SFCL	Superconducting fault current limiter
YBCO	Yttrium barium copper oxide
VOCOL	Voltage-dependent current order limiter

Symbols

I	Current [A]	R	Resistance [Ω]
X	Reactance [Ω]	U	Voltage [V]
P	Power [W]	α	Firing angle [Degree]
M	Modulation ratio [-]	L	Inductance [H]
C	Capacitance [F]		

Subscripts

c	Commutation	f	Fault
T	Transformer	r	Rectifier side
i	Inverter side	g	Ground
ac	Alternating current	dc	Direct current
sm	Smoothing reactor	min	Minimum

Appendix A

Appendix A.1. LCC Station Modeling

Figure A1 shows the connection diagram and equivalent circuit of the LCC station. U_{ac1} and X_{ac1} are the equivalent voltage and reactance of the AC grid coupled to the LCC station, respectively; X_{T1} is the transformer leakage reactance; I_{LCC} and U_{dcr} indicate the equivalent AC current and DC voltage sources of the LCC station.

The switching functions of the AC current and the DC voltage are $S_{LCC\text{-}abci}$ ($S_{LCC\text{-}ai}$, $S_{LCC\text{-}bi}$, $S_{LCC\text{-}ci}$) and $S_{LCC\text{-}abcu}$ ($S_{LCC\text{-}au}$, $S_{LCC\text{-}bu}$, $S_{LCC\text{-}cu}$), respectively. The mathematical equations are obtained as follows:

$$\begin{bmatrix} i_{LCC-a} \\ i_{LCC-b} \\ i_{LCC-c} \end{bmatrix} = \begin{bmatrix} S_{LCC-ai} \\ S_{LCC-bi} \\ S_{LCC-ci} \end{bmatrix} I_{dcr} \tag{A1}$$

$$\begin{bmatrix} U_{LCC-a} \\ U_{LCC-b} \\ U_{LCC-c} \end{bmatrix} = \begin{bmatrix} S_{LCC-au} & S_{LCC-bu} & S_{LCC-cu} \end{bmatrix}^{-1} U_{dcr} \tag{A2}$$

where I_{dcr} is the DC current of the LCC station; U_{LCC} is the voltage over the equivalent AC current source.

Article

Phase Transition and Metallization of Orpiment by Raman Spectroscopy, Electrical Conductivity and Theoretical Calculation under High Pressure

Kaixiang Liu [1,2], Lidong Dai [1,*], Heping Li [1], Haiying Hu [1], Linfei Yang [1,2], Chang Pu [1,2], Meiling Hong [1,2] and Pengfei Liu [3]

[1] Key Laboratory of High-Temperature and High-Pressure Study of the Earth's Interior, Institute of Geochemistry, Chinese Academy of Sciences, Guiyang 550081, China; liukaixiang@mail.gyig.ac.cn (K.L.); hepingli_2007@hotmail.com (H.L.); huhaiying@mail.gyig.ac.cn (H.H.); yanglinfei@mail.gyig.ac.cn (L.Y.); puchang@mail.gyig.ac.cn (C.P.); hongmeilin@mail.gyig.ac.cn (M.H.)
[2] University of Chinese Academy of Sciences, Beijing 100039, China
[3] State Key Laboratory of Structural Chemistry, Fujian Institute of Research on the Structure of Matter, Chinese Academy of Sciences, Fuzhou 350002, China; liupengfei20170208@163.com
* Correspondence: dailidong@gyig.ac.cn

Received: 15 February 2019; Accepted: 5 March 2019; Published: 7 March 2019

Abstract: The structural, vibrational, and electronic characteristics in orpiment were performed in the diamond anvil cell (DAC), combined with a series of experimental and theoretical research, including Raman spectroscopy, impedance spectroscopy, atomic force microscopy (AFM), high-resolution transmission electron microscopy (HRTEM), and first-principles theoretical calculations. The isostructural phase transition at ~25.0 GPa was manifested as noticeable changes in the compressibility, bond lengths, and slope of the conductivity, as well as in a continuous change in the pressure dependence of the unit cell volume. Furthermore, a pressure-induced metallization occurred at ~42.0 GPa, accompanied by reversible electrical conductivity. We also determined the metallicity of orpiment at 45.0 GPa by first-principles theoretical calculations, and the results were in good agreement with the results of the temperature-dependent conductivity measurements. The HRTEM and AFM images of the recovered sample confirmed that orpiment remains in the crystalline phase with an intact layered structure and available crystal-shaped clusters. These high-pressure behaviors of orpiment present some crucial information on the structural phase transition, metallization, amorphization and superconductivity for the A_2B_3-type of engineering materials at high pressure.

Keywords: high pressure; diamond anvil cell; Raman spectroscopy; electrical conductivity; phase transition; pressure-induced metallization

1. Introduction

A_2B_3-type chalcogenides with diverse structures and physical properties could be exploited in some important industrial applications, such as thermoelectric devices, solid-state power devices, refrigerating devices, photovoltaic cells, spintronics, and quantum computation [1–4]. As a representative A_2B_3-type semiconductor, orpiment can be used for photocatalytic water-splitting applications [5]. The amorphous phase of orpiment is also an important material for physical applications because of its extensive use in optics and electronics [6]. Orpiment (As_2S_3) is a well-known binary semiconductor with optical bandgap energy (E_g) of ~2.7 eV [7]. Under ambient conditions, orpiment crystallizes in a quasi-two-dimensional monoclinic structure (SG $P2_1/c$, Z = 4), in which the layers parallel to the (010) plane are bonded by weak van der Waals forces [8].

As we know, pressure is one of the efficient methods to optimize and improve the structural and physical properties in a large number of engineering materials, such as AB, AB_2 and A_2B_3

types of semiconductor compounds [9–12]. As for the representative A_2B_3-type compound, the pressure-induced structural phase transition, metallization, amorphization and superconductivity have already attracted considerable attention by more and more researchers in the recent several years [12–16]. Hence, some crucial characterizations of orpiment under high pressure and high temperature have been reported previously. Bolotina et al. [17] observed decomposition of orpiment into two high-pressure phases (AsS and AsS_2) at a pressure above ~6 GPa and a temperature above ~800 K, by means of Xcalibur single-crystal diffraction. Besson et al. [18] found that when the pressure increased to 10 GPa at room temperature, there was no phase transition of orpiment based on optical absorption and Raman spectroscopy experiments, even though there was a decrease in the bandgap energy from 2.7 eV to 1.6 eV. Although reports about the experimental data for orpiment at pressures above 10 GPa are absent, high pressure can be applied to reduce the interatomic distances, finally resulting in metallization and a possible pressure-induced phase transition. For amorphous orpiment, from Raman spectroscopy measurements, a pressure-induced phase transition occurred at ~4 GPa [19,20]. In addition, the semiconductor–metal transition of amorphous orpiment has been observed at a pressure of ~45 GPa by optical reflectivity and absorption measurements [21].

The type of phase transition can be determined by the characteristic parameter variations, such as variation in the unit cell volume, crystalline lattice parameters, axial ratio and bond lengths [22,23]. These characteristic parameters can be obtained by first-principles theoretical calculations. Radescu et al. [24] recently determined that there was no phase transition in orpiment up to 16 GPa, by first-principles theoretical calculations. To verify whether orpiment undergoes a pressure-induced phase transition above 16 GPa, the characteristic crystal cell parameters should be calculated in a large pressure range.

In the present study, to systematically investigate the pressure-induced phase transition and metallization of orpiment, we determined the electrical and structural properties at pressures up to ~46.0 GPa using the DAC in conjunction with a series of experimental and theoretical methods, including Raman spectroscopy, impedance spectroscopy, atomic force microscopy (AFM), high-resolution transmission electron microscopy (HRTEM), and first-principles theoretical calculations.

2. Experimental and Computational Details

2.1. Sample Description

In this study, the natural crystalline orpiment was gathered from Jiepaiyu ore deposit in Shimen city, Hunan province. Before the high-pressure experiments, the crystalline orpiment was crushed into a powder (~20 μm). The X-ray powder diffraction (XRD) analysis of the sample was conducted using an X'Pert Pro X-ray powder diffractometer (Phillips Company, Amsterdam, Netherlands), the Cu Kα radiation with 45 kV and 40 mA) in the State Key Laboratory of Ore Deposit Geochemistry, Institute of Geochemistry, Chinese Academy of Sciences. Figure 1 is the X-ray diffraction for orpiment under ambient conditions. The samples displayed a quasi-two-dimensional monoclinic structure (SG $P2_1/c$, $Z = 4$). The data analysis and handling software JADE 6.0 was used. Some lattice constant parameters were given as follows: $a = 4.22$ Å, $b = 9.57$ Å, $c = 11.46$ Å and $\beta = 90.5°$. The unit cell volume (V) was 462.8 Å^3.

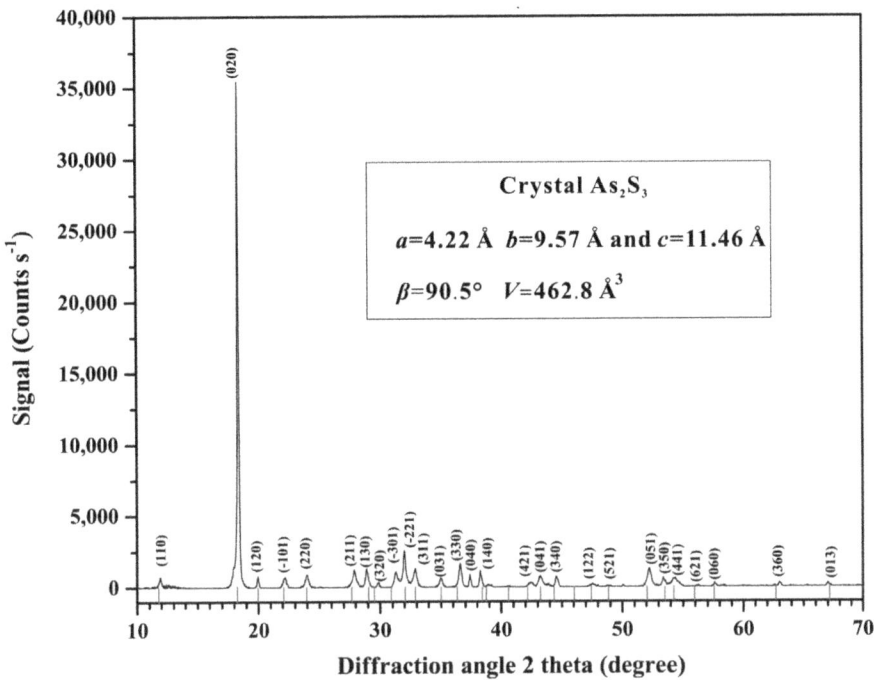

Figure 1. The X-ray diffraction of orpiment under ambient conditions.

2.2. High-Pressure Raman Scattering Measurements

A diamond anvil cell (DAC) with a 300 μm anvil culet was adopted in the Raman spectroscopy measurements (Renishaw, London, England) under high pressure. The pressure calibration was realized using the ruby luminescence method. Helium was used as the pressure medium to provide a hydrostatic condition, and no pressure medium was used for the nonhydrostatic condition. The Raman spectra were recorded with an Invia Raman spectrometer (Renishaw, London, England) equipped with a charge-coupled device camera (Olympus, Tokyo, Japan) and a confocal microscope (TCS SP8, Leica, Solms, Germany). The excitation laser power for the high-pressure Raman spectra measurements and fluorescence was 20 mW and 0.5–40 μW, respectively. The Raman spectra were obtained by an argon ion laser (Spectra physics; 514.5 nm, power <1 mW) in the backscattering geometry, and a Raman range of 125–425 cm^{-1} with the resolution of 1.0 cm^{-1} was employed in the process of spectral acquisition. In order to achieve a stable pressure condition, the equipment pressure stabilization time was 1 h at the predesigned pressure, before each Raman spectral measurement. The PeakFit software was employed to fit the correspondent Raman spectroscopy so as to identify the position of each Raman mode and its uncertainty. The AFM and TEM data were measured by virtue of a Multimode 8 mass spectrometer (Bruker, Karlsruhe, Germany) and a Tecnai G2 F20 S-TWIN TMP (FEI, Hillsboro, America), respectively.

2.3. High-Pressure Conductivity Measurements

For the high-pressure electrical conductivity experiments, a DAC with a 300 μm anvil culet was used. After pre-indented to a thickness of ~60 μm, a 180 μm hole was drilled in a T-301 gasket using a laser. Then the hole was filled with the insulating powder, which consisted of a boron nitride powder and epoxy, and another 100 μm hole was drilled as the insulating sample chamber. Figure 2 is the cross-sectional structure of the DAC. The electrical conductivity of orpiment was acquired using Solartron-1260 and Solartron-1296 impedance spectroscopy analyzers, with a frequency of 10^{-1}–10^7 Hz. For the temperature-dependent conductivity measurements, liquid nitrogen was used to obtain different temperatures. The temperature measurements were performed by a k-type thermocouple, which was attached to the diamond with an accuracy of 5 K. The temperature of the experimental assembly was varied by volatilization of liquid nitrogen. A similar measurement procedure and experimental assembly were presented previously [12,25–27].

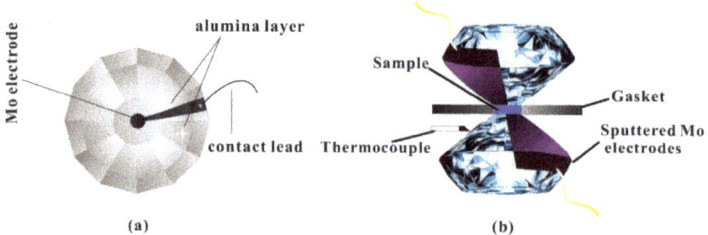

Figure 2. Measurement assemblage of sample for electrical conductivity at high pressure. (**a**) The structure of plate electrodes integrated on two diamond anvils. (**b**) Cross section of the diamond anvil cell (DAC) employed in the high-pressure electrical conductivity measurement.

2.4. Computational Details

All of the ab initio calculations were performed with the CASTEP (Materials Studio) code within the first-principles theoretical framework of density functional theory (DFT), using the pseudopotential method. The Perdew–Burke–Ernzerhof scheme in generalized gradient approximation was used to obtain the exchange and correlation terms. The Broyden–Fletcher–Goldfarb–Shanno minimization algorithm in the code was adopted to realize structural optimizations in orpiment. A cutoff energy of 360 eV was applied to the valence electronic wave functions expanded in a plane-wave basis set for monoclinic orpiment. To guarantee the high convergence in the total energy of 1 meV per atom, the special k points generated by $6 \times 4 \times 2$ parameter grids for the $P2_1/c$ phase were acquired to achieve the integration of the Brillouin zone. All of the atoms and lattice constants were relaxed thoroughly until the force convergence reduced to less than 0.01 eV/Å, which was used to obtain the different lattice constants and atomic positions of orpiment at varied pressures. The initial structural parameters of orpiment were obtained from previously reported results [28].

3. Results and Discussion

In this work, the Raman scattering experiments were conducted under non-hydrostatic and hydrostatic conditions at a pressure range of 1 atm to 42.6 GPa. A series of non-hydrostatic Raman scattering peaks and their corresponding Raman shift results are shown in Figure 3. Another similar hydrostatic Raman peak and its corresponding Raman shift data are given in Figure 4. As shown in Figure 3a, nine characteristic Raman active modes were observed at an ambient pressure, which were assigned as follows [29,30]. The peaks at 135 and 201 cm^{-1} stand for the As–S–As bending vibration. The peak at 153 cm^{-1} denotes to the As–As–S bending vibration. The peaks at 177 and 187 cm^{-1} were assigned to S–As–S bending vibrations. The peaks at 290 and 308 cm^{-1} denote antisymmetric As–S stretching vibrations. The peak at 353 cm^{-1} denotes the As–S stretching vibration. The peak at 380 cm^{-1} was assigned to the antisymmetric As–S–As stretching vibration. These obtained nine Raman vibration modes were consistent with the results reported by Cheng et al [29]. The Raman-active modes of orpiment continuously shifted towards higher frequencies with increasing pressure, except for the Raman mode with a peak position of 290 cm^{-1}. When the pressure increased to above 6.5 GPa, some remarkable characteristics were detected in the Raman spectrum of orpiment. The modes with peak positions of 135 and 177 cm^{-1} merged with their neighboring modes at 6.5 and 9.1 GPa, respectively. Three new Raman-active modes appeared: A peak at 145 cm^{-1} at the pressure of 6.5 GPa; and peaks at 347 and 365 cm^{-1} at 7.6 GPa. The newly appeared Raman mode with a peak position of 145 cm^{-1} shifted to a lower frequency with increasing pressure, while the Raman modes with peak positions of 347 and 365 cm^{-1} shifted to higher frequencies. After the pressure was enhanced to 17.0 GPa, the Raman features tended to be difficult to distinguish, until they disappear. The Raman spectrum of orpiment upon decompression from 42.6 GPa recovered to the original state, which meant a reversible process under a non-hydrostatic condition.

The evolution of the experimental Raman modes in orpiment with increasing pressure is shown in Figure 3b. All of these Raman modes at ambient pressure showed nonlinear behavior with increasing pressure under our experimental conditions. The obtained fitting results and the pressure coefficients are summaries in Table 1, using the equation:

$$\omega(P) = \omega_0 + \alpha P + \beta P^2 \tag{1}$$

where ω_0 is the peak positions of the Raman modes at an ambient condition and P is the pressure. It was remarkable that several Raman vibration modes of orpiment exhibited moderate softening, merged with their neighbor modes and exhibited a complex splitting behavior in a given pressure range. The new appeared Raman modes above 6.0 GPa with peaks at 145, 347 and 365 cm^{-1} were also detected by Mamandov et al. [31] at ambient pressure and a low temperature (T = 4 K). This phenomenon illustrated the complementary aspect of a low temperature (at P = 0) and a high pressure (at T = 300 K) for Raman spectroscopy: A low temperature reduced the line width to resolve as many structural features as possible for these complex spectra, while a high pressure revealed the structure by increased splitting at a constant line width [18]. All the obtained experimental points did not show obvious unusual behavior at any given pressure point. Indeed, the obtained relationship between the observed Raman modes and the pressure could be attributed to the variations of the atomic positions and bond lengths at high pressure. A similar trend existed for the hydrostatic conditions shown in Figure 4. Consequently, the results of Raman spectroscopy measurements at high pressure disclosed good phase stability of orpiment below 17.0 GPa.

Table 1. Relationship between the pressure and the Raman shift for orpiment as fitted with equation: $\omega(P) = \omega_0 + \alpha P + \beta P^2$.

Mode Number	ω_0 (cm^{-1})	α (cm^{-1}GPa^{-1})	β (cm^{-1}GPa^{-2})
1	135	−0.420	0.120
2	153	0.240	0.095
3	177	−2.220	0.320
4	187	0.683	0.013
5	201	0.125	0.147
6	290	−0.891	0.046
7	308	−0.204	0.123
8	353	−0.193	0.118
9	380	0.474	0.107

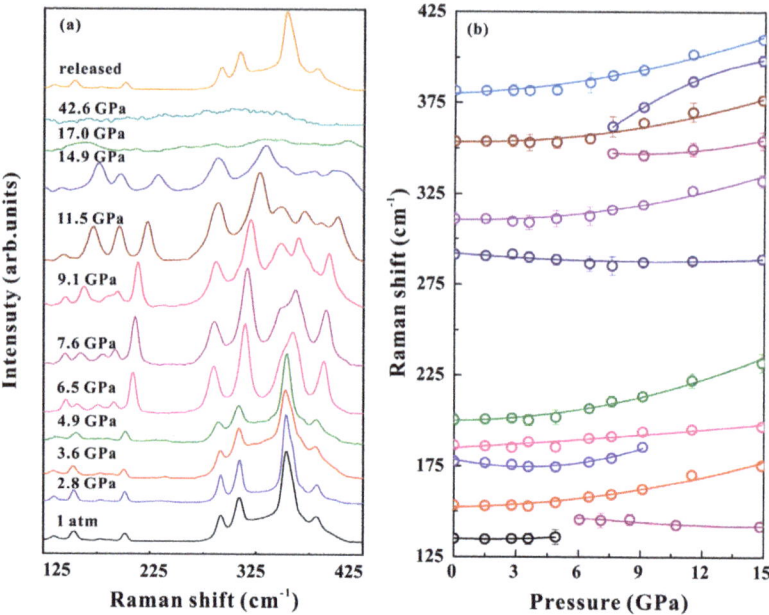

Figure 3. (a) Raman spectra of orpiment at selected pressures under non-hydrostatic condition (λ = 514 nm, T = 300 K). (b) Raman shift of orpiment with increasing pressure.

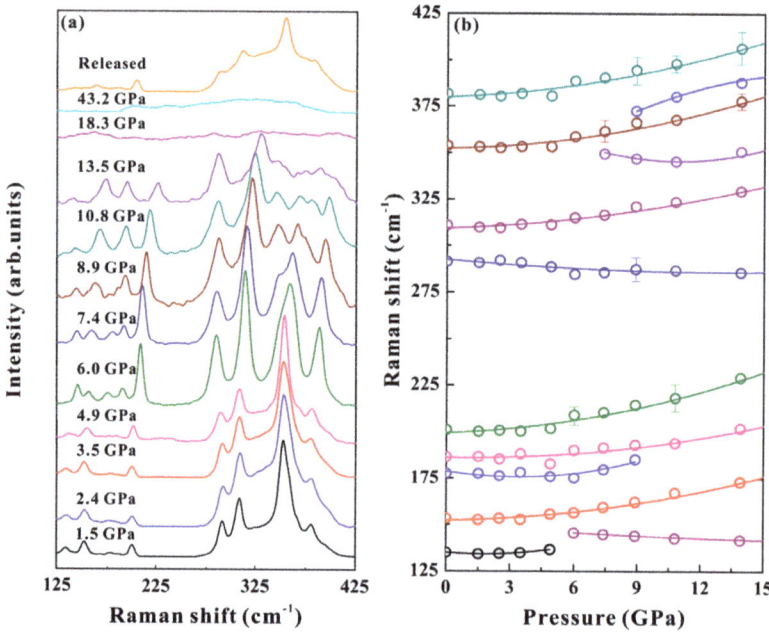

Figure 4. (a) Raman spectra of orpiment at selected pressures under hydrostatic condition (λ = 514 nm, T = 300 K). (b) Raman mode frequency evolution against pressure.

Both the variations in the distances of atoms and the structural phase transition in orpiment could tune its electronic properties. To further verify the phase transition and metallization of orpiment, electrical conductivity measurements were performed up to ~44.0 GPa at room temperature. The representative complex impedance spectra of orpiment under high pressure are shown in Figure 5a–c. The ZView software was used to fit the plots (equivalent circuit method). Two parts could be well identified in the frequency range 10^{-1}–10^7 of the impedance spectra: The semicircular arc in the higher frequency represented the resistance of the grain interior, whereas the oblique line at a lower frequency was characteristic of the grain boundary [32,33]. It is noteworthy that the resistance of grain boundary began to decrease above 24.4 GPa, and disappeared when the pressure was enhanced to 34.0 GPa. In this work, we paid more attention to the pressure effects of the grain interior contribution related to the phase transitions. The relationships between the electrical conductivity of orpiment and pressure in the process of compression and decompression at room temperature are shown in Figure 5d. As the pressure increased, the electrical conductivity of orpiment decreased before 24.4 GPa, and it then increased rapidly up to the pressure of 38.2 GPa. When the pressure was enhanced to 41.2 GPa, the electrical conductivity of orpiment remained relatively stable. The pressure-dependent electrical conductivity of orpiment in our experimental pressure range could be divided into three parts: An ambient pressure to 24.4 GPa with a rate of -0.020 S cm^{-1}, 24.4 to 38.2 GPa with a rate of 0.162 S cm^{-1}, and 38.2 to 44.0 GPa with a rate of 0.018 S cm^{-1}. The variation in the conductivity of orpiment at 24.4 GPa was related to a pressure-induced phase transition. The increased overlap of the electronic orbital wave function and narrowing of the energy gap are reasons why the electrical conductivity of orpiment showed a rapidly increasing trend between 24.4 and 38.2 GPa (Figure 5d). Electrical conductivity of greater than 1 S cm^{-1} at a pressure above 38.2 GPa may be indicative of metallization. The electrical conductivity of orpiment was reversible upon decompression, which was consistent with the Raman spectroscopy measurements. This reversible phenomenon on the electrical conductivity of orpiment was different with Sb_2S_3, with a layer structure [12].

Temperature-dependent conductivity measurements of orpiment were performed to verify whether orpiment undergoes metallization. The results can be fitted by the Arrhenius equation. With increasing temperature, the electrical conductivity of orpiment increased below 41.0 GPa, which represented typical semiconductor behavior. It showed a negative relationship between the temperature and electrical conductivity at 42.5 GPa, which indicated a clear metallic behavior (Figure 6b). All of the obtained results revealed the occurrence of the semiconductor-metal transition of orpiment. According to the relationship between the temperature and electrical conductivity, the activation energy of orpiment at a selected pressure can be determined by:

$$\sigma = \sigma_0 \exp(-E_t/k_b T) \qquad (2)$$

where σ_0 stands for the pre-exponential factor (S cm^{-1}), E_t stands for the activation energy (meV)—which could be determined by a linear fitting between the logarithmic conductivity and $1000/T$—k_b stands for the Boltzmann constant, and T stands for the absolute temperature (K). The relationship between the activation energy of orpiment and pressure is shown in Figure 6c. The activation energy reduced with increasing pressure, which indicated that electrical transport of carriers became easier at high pressure. The pressure dependence of the activation energy can be fitted as a function of:

$$E_t = -3.28 + 137.69P \qquad (3)$$

where E_t denotes the activation energy and P denotes the pressure. The fitting results demonstrated that the value of E_t reached zero when the pressure was enhanced up to 41.9 GPa, as shown in Figure 6c. This meant that this sample became metal above 41.9 GPa, which was in good agreement with the electrical conductivity measured results.

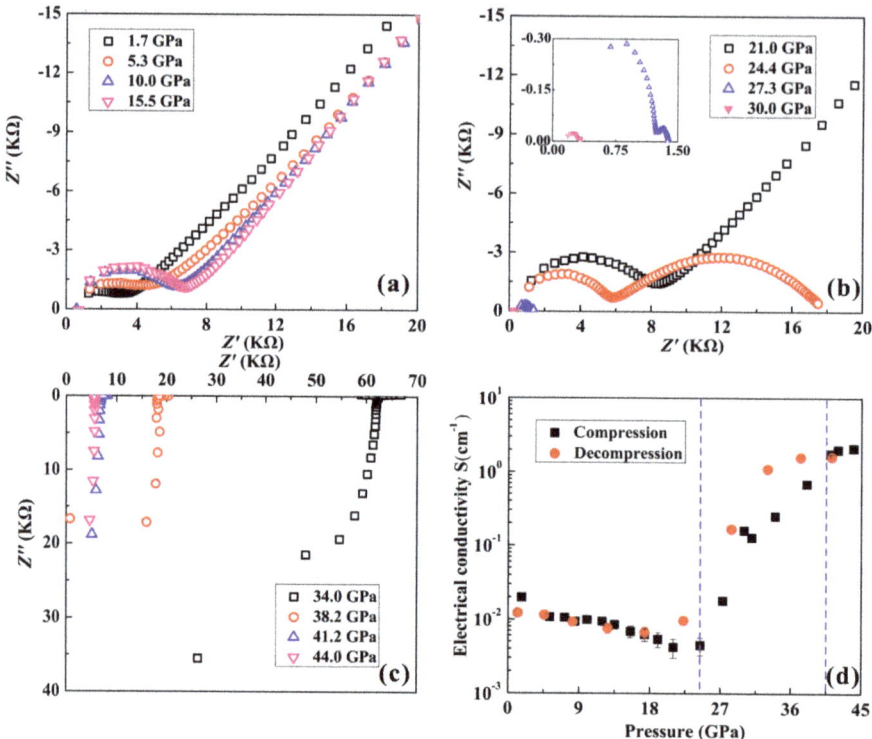

Figure 5. The electrical behaviors for orpiment with increasing pressure. (**a**–**c**) The complex impedance spectra of orpiment at selected pressures. (**d**) The variations of electrical conductivity for orpiment with the increasing pressure and decreasing pressure.

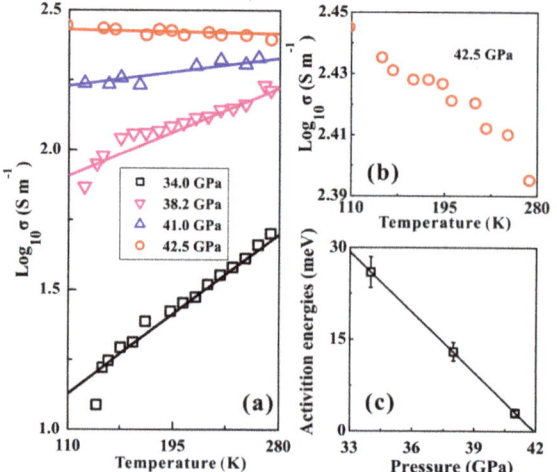

Figure 6. (**a**) The relationship between the temperature and electrical of orpiment at selected pressure. (**b**) Pressure dependence of activation energy for orpiment.

To deeply explore the morphology and structural changes of orpiment after decompression from 44.0 GPa, the analysis for the recovered orpiment was performed by HRTEM and AFM (Figure 7). As shown in Figure 7a, the interlayer spacing of the starting material was ~0.45 nm. As for the recovered orpiment released from 44.0 GPa, it remained in the crystalline phase with an interlayer spacing of ~0.34 nm (Figure 7a). As shown in Figure 7b, there were crystal-shaped clusters on the recovered samples. Both the layer structure and surface morphology were well preserved upon decompression from 44.0 GPa. This reversible phenomenon was consistent with the results of the electrical conductivity and Raman spectroscopy measurements.

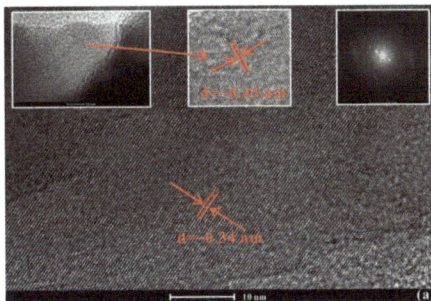 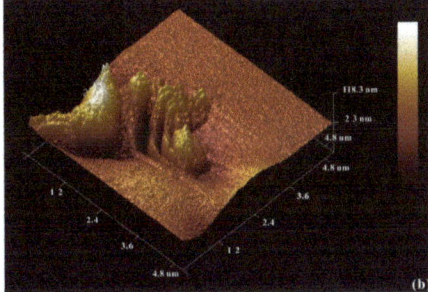

Figure 7. (a) The high-resolution transmission electron microscopy (HRTEM) images of decompressed orpiment from 44.0 GPa. Inset: the left one is a HRTEM image of the initial sample, the right one is a cross-sectional selected-area electron diffraction pattern by HRTEM for the decompressed orpiment. (b) The atomic force microscopy (AFM) image of orpiment upon decompression from 44.0 GPa.

To further investigate the phase stability of orpiment under high pressure, first-principles theoretical calculations were also performed at a pressure range from 0 GPa to 46.0 GPa. Some crucial crystalline parameters of orpiment, including the unit cell volume, lattice parameters, and bond angles were determined, as shown in Figure 8a–c. The obtained value of unit cell volume was 471.2 Å^3 and the lattice parameters were a = 4.22 Å, b = 9.65 Å, c = 12.27 Å, and β = 109.59°, respectively, which was consistent with previously reported results [24,28]. The pressure dependence of the lattice parameters is shown in Figure 8a, from which the anisotropic compressibility of orpiment could be determined. In particular, the short a axis had better compressibility than the b and c axes, with increasing pressure. The stiffer lattice constant of the long c axis decreased with the increasing pressure, which showed a simple monotonic liner behavior with increasing pressure. By taking into account the lattice constant ratios, some interesting variations of pressure effects in both a/c and b/c were observed at ~25 GPa (Figure 8b). The compressibility of the initially softer a and b axes reach that of the c axis above ~25 GPa. The variations of axial compressibility were directly related to changes in interatomic parameters. There was a continuous change in the pressure dependence of the unit cell volume up to 46.0 GPa (Figure 8c). The pressure dependence of selected interatomic As–S bond lengths of orpiment at high pressure is shown in Figure 8e. The most significant stretching of the bond in the calculation below ~25 GPa, was the S3–As1 bond, while the S1–As1 bond showed the least significant stretching. For the S2–As2 and S3–As2 bonds, similar stretching was observed below ~25 GPa. The selected bond length showed the most obvious pressure-related effect, which underwent a reversal in the pressure dependence at ~25 GPa. The changes of the pressure effects in the bond lengths were linked to the compressibility changes of the axes, as mentioned above.

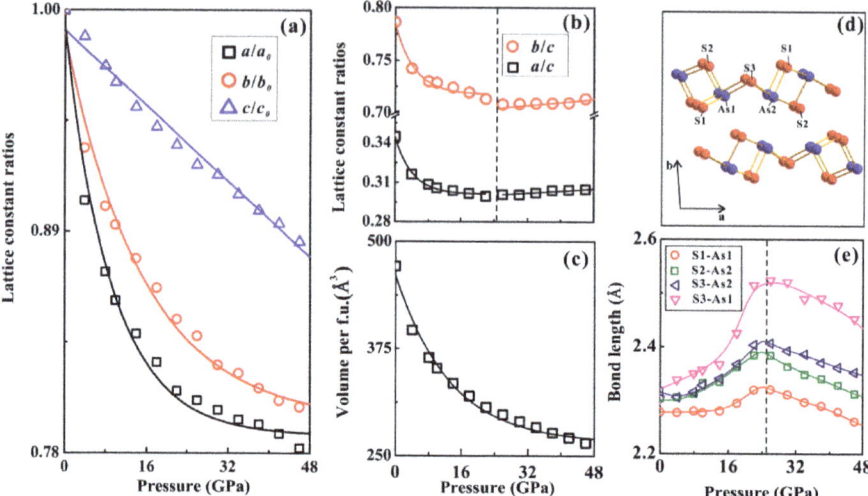

Figure 8. The variations of (**a**) lattice constant parameters (a/a_0, b/b_0 and c/c_0), (**b**) axial ratios (a/c and b/c), (**c**) unit cell volume and bond length for orpiment at high pressure. Two regions could be determined by the black vertical dotted lines at the pressure of ~25 GPa (**d**) Crystalline structure of monoclinic ($P2_1/c$) orpiment at ambient pressure. We also nominate the As and S ions in the crystalline structure. (**e**) Selected S-As bond lengths of orpiment up to 46 GPa.

The type of phase transition can be determined by the characteristic parameter variations, such as the variations in the unit cell volume, lattice parameters, axial ratios, and bond lengths [22,23]. The collapse of a unit cell volume provides good evidence for the structural phase transition under high pressure. Whereas, the observed discontinuities in the axial ratios and bond lengths with the continuities in the unit cell volume and lattice parameter at a certain pressure are the characters for the isostructural phase transition. Thus, we attributed the variations in compressibility of orpiment at ~25.0 GPa to an isostructural phase transition.

At the same time, the evolution of the electronic properties of orpiment at high pressure was also revealed through first-principles theoretical calculations. The electrical structures of orpiment at different pressures are shown in Figure 9. These results predicted that orpiment had indirect bandgap energy of 2.07 eV under ambient conditions, which agreed well with a previous study [5]. The bandgap energy decreased to 0.51 eV at 20.0 GPa and then closed at 45.0 GPa, which indicated a clear metallicity of orpiment. These results provided more evidence for pressure-induced metallization of orpiment at around 42.0 GPa, which was consistent with the experimental results. As seen in Figure 9, the high-energy valence bands and the conduction band were controlled by the S–p and As–p states, respectively. The low-energy valence bands were contributed mostly by the S-s states, whereas, the middle-energy valence bands were contributed mostly by the As-s states. Electronic coupling and hybridization became gradually intense at a high pressure, which resulted in the broadening of the energy bands. Furthermore, compared with the conduction band, the high-energy valence band of orpiment showed a stronger broadening with increasing pressure, which could result in a decrease in the bandgap and even closing of the bandgap.

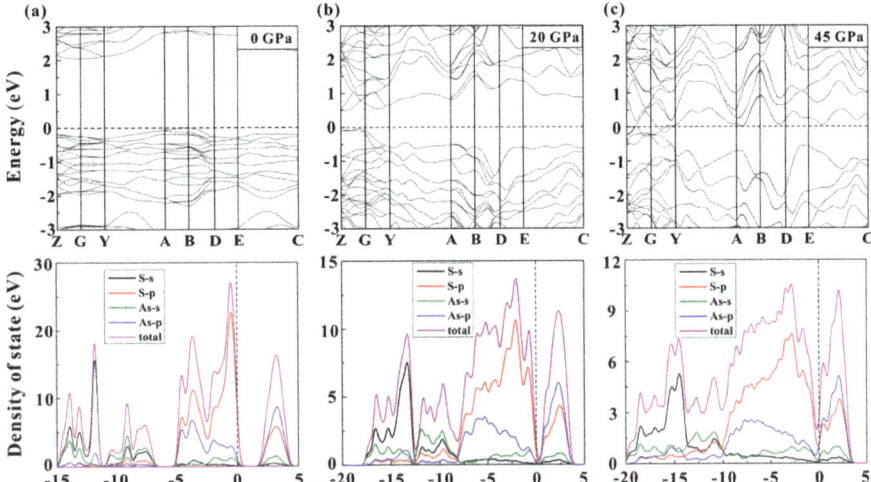

Figure 9. The selected band structure and the corresponding total and partial density for orpiment at different pressure. (**a**) and (**b**) The bandgap energy at 0 and 20 GPa are 2.07 and 0.51 eV, respectively. (**c**) The bandgap has already closed at the pressure of 45 GPa. The bandgap of orpiment narrows with increasing pressure.

4. Conclusions

Using a DAC, we have found that the phase transition and metallization of orpiment occurred at about 25.0 and 42.0 GPa, respectively. The results were acquired by a combination of experimental and theoretical methods, including Raman scattering, impedance spectroscopy, electrical conductivity measurements at variable temperature, AFM, HRTEM, and first-principles calculations. The variable temperature electrical conductivity measurements and first-principles theoretical calculations provided strong evidence for a pressure-induced metallization at ~42.0 GPa. The images of the decompressed sample, from AFM and HRTEM, confirmed the well-preserved crystalline structure, which agreed well with the electrical conductivity and Raman spectroscopy measurements. Based on the observed compressibility change and the variation in pressure effects on the electrical conductivity, a second-order isostructural phase transition occurred at ~25.0 GPa. The observed high-pressure properties of orpiment will aid in the understanding of the universal crystal structure and electrical properties of A_2B_3-type materials. And furthermore, all of these high-pressure behaviors of orpiment will present some crucial information in the structural phase transition, metallization, amorphization and superconductivity for the A_2B_3-type of engineering materials at high pressure.

Author Contributions: L.D. (designing the project). K.L. and L.D. (writing the initial draft of the work and the final paper). K.L., L.D., H.L., H.H., L.Y., C.P., M.H. and P.L. (interpreting the results). L.D. (correcting and recognizing the final paper). K.L. and L.Y. (performing and interpreting the high-P experiments, the HRTEM and AFM images). K.L. and P.L. (performing the first-principles calculations). All authors discussed the results and commented on the manuscript.

Funding: This research was financially supported by the strategic priority Research Program (B) of the Chinese Academy of Sciences (Grant No. 18010401), Key Research Program of Frontier Sciences of CAS (Grant No. QYZDB-SSW-DQC009), "135" Program of the Institute of Geochemistry of CAS, Hundred Talents Program of CAS and NSF of China (Grant Nos. 41474078, 41774099 and 41772042).

Acknowledgments: The support of the Supercomputer Center of Fujian Institute of Research on the Structure of Matter (FJIRSM) is acknowledged.

Conflicts of Interest: The authors declare no conflict of interest.

References

1. Snyder, G.J.; Toberer, E.S. Complex thermoelectric materials. *Nat. Mater.* **2008**, *7*, 105. [CrossRef] [PubMed]
2. Pawar, S.H.; Bhosale, P.N.; Uplane, M.D.; Tamhankar, S. Growth of Bi_2S_3 film using a solution-gas interface technique. *Thin Solid Films* **1983**, *110*, 165–170. [CrossRef]
3. Hasan, M.Z.; Kane, C.L. Colloquium: Topological insulators. *Rev. Mod. Phys.* **2010**, *82*, 3045–3067. [CrossRef]
4. Moore, J.E. The birth of topological insulators. *Nature* **2010**, *464*, 194–198. [CrossRef] [PubMed]
5. Debbichi, L.; Kim, H.; Bjorkman, T.; Eriksson, O.; Lebegue, S. First-principles investigation of two-dimensional trichalcogenide and sesquichalcogenide monolayers. *Phys. Rev. B: Condens. Matter* **2016**, *93*, 245307. [CrossRef]
6. Asobe, M. Nonlinear optical properties of chalcogenide glass fibers and their application to all-optical switching. *Opt. Fiber Technol.* **1997**, *3*, 142–148. [CrossRef]
7. Zallen, R. Effect of pressure on optical properties of crystalline As_2S_3. *High Pressure Res.* **2004**, *24*, 117–118. [CrossRef]
8. Lavrentiev, A.A.; Gabrel'yan, B.V.; Nikiforov, I.Y.; Vorzhev, V.B. Electronic energy structure of As_2S_3, AsSI, $AgAsS_2$, and TiS_2 semiconductors. *J. Struct. Chem.* **2005**, *46*, 805–812. [CrossRef]
9. Pu, C.; Dai, L.D.; Li, H.P.; Hu, H.Y.; Liu, K.X.; Yang, L.F.; Hong, M.L. Pressure-induced phase transitions of ZnSe under different pressure environments. *AIP Adv.* **2019**, *9*, 025004. [CrossRef]
10. Yang, L.F.; Dai, L.D.; Li, H.P.; Hu, H.Y.; Liu, K.X.; Pu, C.; Hong, M.L. Pressure-induced metallization in $MoSe_2$ under different pressure conditions. *RSC Adv.* **2019**, *9*, 5794–5803. [CrossRef]
11. Dai, L.D.; Zhuang, Y.K.; Li, H.P.; Wu, L.; Hu, H.Y.; Liu, K.X.; Yang, L.F.; Pu, C. Pressure-induced irreversible amorphization and metallization with a structural phase transition in arsenic telluride. *J. Mater. Chem.* **2017**, *5*, 12157–12162. [CrossRef]
12. Dai, L.D.; Liu, K.X.; Li, H.P.; Wu, L.; Hu, H.Y.; Zhuang, Y.K.; Yang, L.F.; Pu, C.; Liu, P.F. Pressure-induced irreversible metallization accompanying the phase transitions in Sb_2S_3. *Phys. Rev. B: Condens. Matter* **2018**, *97*, 024103. [CrossRef]
13. Vilaplana, R.; Gomis, O.; Manjón, F.J.; Segura, A.; Pérez-González, E.; Rodríguez-Hernández, P.; Muñoz, A.; González, J.; Marín-Borrás, V.; Muñoz-Sanjosé, V.; et al. High-pressure vibrational and optical study of Bi_2Te_3. *Phys. Rev. B: Condens. Matter* **2011**, *84*, 104112. [CrossRef]
14. Zhang, J.L.; Zhang, S.J.; Weng, H.M.; Zhang, W.; Yang, L.X.; Liu, Q.Q.; Feng, S.M.; Wang, X.C.; Yu, R.C.; Cao, L.Z.; et al. Pressure-induced superconductivity in topological parent compound Bi_2Te_3. *Proc. Natl. Acad. Sci. U.S.A.* **2011**, *108*, 24. [CrossRef] [PubMed]
15. Zhang, J.K.; Liu, C.L.; Zhang, X.; Ke, F.; Han, Y.H.; Peng, G.; Ma, Y.Z.; Gao, C.X. Electronic topological transition and semiconductor-to-metal conversion of Bi_2Te_3 under high pressure. *Appl. Phys. Lett.* **2013**, *103*, 052102. [CrossRef]
16. Kong, P.P.; Sun, F.; Xing, L.Y.; Zhu, J.; Zhang, S.J.; Li, W.M.; Liu, Q.Q.; Wang, X.C.; Feng, S.M.; Yu, X.H.; et al. Superconductivity in strong spin orbital coupling compound Sb_2Se_3. *Sci. Rep.* **2014**, *4*, 6679. [CrossRef] [PubMed]
17. Bolotina, N.B.; Brazhkin, V.V.; Dyuzheva, T.I.; Katayama, Y.; Kulikova, L.F.; Lityagina, L.V.; Nikolaev, N.A. High-pressure polymorphism of As_2S_3 and new AsS_2 modification with layered structure. *JETP. Lett.* **2013**, *98*, 539–543. [CrossRef]
18. Besson, J.M.; Cernogora, J.; Zallen, R. Effect of pressure on optical properties of crystalline As_2S_3. *Phys. Rev. B: Condens. Matter* **1980**, *22*, 3866. [CrossRef]
19. Andrikopoulos, K.S.; Christofilos, D.; Kourouklis, G.A.; Yannopoulos, S.N. Pressure Raman study of vibrational modes of glassy As_2X_3 (X:O, S). *High Pressure Res.* **2006**, *26*, 401–406. [CrossRef]
20. Andrikopoulos, K.S.; Christofilos, D.; Kourouklis, G.A.; Yannopoulos, S.N. Pressure dependence of the boson peak in glassy As_2S_3 studied by Raman scattering. *J. Non-Cryst. Solids* **2006**, *352*, 4594–4600. [CrossRef]
21. Struzhkin, V.V.; Goncharov, A.F.; Caracas, R.; Mao, H.K.; Hemley, R.J. Synchrotron infrared spectroscopy of the pressure-induced insulator-metal transitions in glassy As_2S_3 and As_2Se_3. *Phys. Rev. B: Condens. Matter* **2008**, *77*, 165133. [CrossRef]
22. Efthimiopoulos, I.; Kemichick, J.; Zhou, X.; Khare, S.V.; Ikuta, D.; Wang, Y.J. High-pressure studies of Bi_2S_3. *J. Phys. Chem. A* **2014**, *118*, 1713–1720. [CrossRef] [PubMed]

23. Zhao, J.G.; Yang, L.X.; Yu, Z.H.; Wang, Y.C.; Li, Y.; Yang, K.; Liu, Z.G.; Wang, Y. Structural phase transitions and metallized phenomena in arsenic telluride under High Pressure. *Inorg. Chem.* **2016**, *55*, 3907–3914. [CrossRef] [PubMed]
24. Radescu, S.; Mujica1, A.; Rodrguez Hernandez, P.; Munoz, A.; Ibanez, J.; Sans, J.A.; Cuenca-Gotor, V.P.; Manjon, F.J. Study of the orpiment and anorpiment phases of As_2S_3 under pressure. *J. Phys. Conf. Ser.* **2017**, *950*, 042018. [CrossRef]
25. Liu, K.X.; Dai, L.D.; Li, H.P.; Wu, L.; Hu, H.Y.; Zhuang, Y.K.; Yang, L.F.; Pu, C. Migration of impurity level reflected in the electrical conductivity variation for natural pyrite at high temperature and high pressure. *Phys. Chem. Miner.* **2018**, *45*, 85–92. [CrossRef]
26. Liu, K.X.; Dai, L.D.; Li, H.P.; Wu, L.; Hu, H.Y.; Zhuang, Y.K.; Yang, L.F.; Pu, C.; Hong, M.L. Pressure-induced phase transitions for goethite investigated by Raman spectroscopy and electrical conductivity. *High Pressure Res.* **2019**, *38*, 106–116. [CrossRef]
27. Yang, L.F.; Dai, L.D.; Li, H.P.; Hu, H.Y.; Zhuang, Y.K.; Liu, K.X.; Pu, C.; Hong, M.L. Pressure-induced structural phase transition and dehydration for gypsum investigated by Raman spectroscopy and electrical conductivity. *Chem. Phys. Lett.* **2018**, *706*, 151–157. [CrossRef]
28. Gibbs, G.V.; Wallace, A.F.; Zallen, R.R.; Downs, T.; Ross, N.L.; Cox, D.F.; Rosso, K.M. Bond paths and van der Waals interactions in orpiment, As_2S_3. *J. Phys. Chem.* **2010**, *114*, 6550–6557. [CrossRef] [PubMed]
29. Cheng, H.F.; Zhou, Y.; Frost, R.L. Structure comparison of orpiment and realgar by Raman spectroscopy. *Spectrosc. Lett.* **2017**, *50*, 23–29. [CrossRef]
30. Roberto, F. The infrared and Raman spectra of realgar and orpiment. *Am. Mineral.* **1969**, *54*, 1062–1074.
31. Mamedov, S.; Drichko, N. Characterization of 2D As_2S_3 crystal by Raman spectroscopy. *MRS. Adv.* **2018**, *1*, 385–390. [CrossRef]
32. Dai, L.D.; Hu, H.Y.; Li, H.P.; Wu, L.; Hui, K.S.; Jiang, J.J.; Sun, W.Q. Influence of temperature, pressure, and oxygen fugacity on the electrical conductivity of dry eclogite, and geophysical implications. *Geochem. Geophys. Geosyst.* **2016**, *17*, 2394–2407. [CrossRef]
33. Hu, H.Y.; Dai, L.D.; Li, H.P.; Hui, K.S.; Sun, W.Q. Influence of dehydration on the electrical conductivity of epidote and implications for high-conductivity anomalies in subduction zones. *J. Geophys. Res. Solid Earth* **2017**, *122*, 2751–2762. [CrossRef]

 © 2019 by the authors. Licensee MDPI, Basel, Switzerland. This article is an open access article distributed under the terms and conditions of the Creative Commons Attribution (CC BY) license (http://creativecommons.org/licenses/by/4.0/).

Article

Microstructure and Flux Pinning of Reacted-and-Pressed, Polycrystalline $Ba_{0.6}K_{0.4}Fe_2As_2$ Powders

Michael R. Koblischka [1,2,*], Anjela Koblischka-Veneva [1,2], Jörg Schmauch [1] and Masato Murakami [2]

1. Experimental Physics, Saarland University, P.O. Box 151150, D-66044 Saarbrücken, Germany
2. Superconducting Materials Laboratory, Department of Materials Science and Engineering, Shibaura Institute of Technology, Tokyo 135-8548, Japan
* Correspondence: m.koblischka@gmail.com or miko@shibaura-it.ac.jp

Received: 7 June 2019; Accepted: 2 July 2019; Published: 6 July 2019

Abstract: The flux pinning properties of reacted-and-pressed $Ba_{0.6}K_{0.4}Fe_2As_2$ powder were measured using magnetic hysteresis loops in the temperature range 20 K $\leq T \leq$ 35 K. The scaling analysis of the flux pinning forces ($F_p = j_c \times B$, with j_c denoting the critical current density) following the Dew-Hughes model reveals a dominant flux pinning provided by normal-conducting point defects (δl-pinning) with only small irreversibility fields, H_{irr}, ranging between 0.5 T (35 K) and 16 T (20 K). Kramer plots demonstrate a linear behavior above an applied field of 0.6 T. The samples were further characterized by electron backscatter diffraction (EBSD) analysis to elucidate the origin of the flux pinning. We compare our data with results of Weiss et al. (bulks) and Yao et al. (tapes), revealing that the dominant flux pinning in the samples for applications is provided mainly by grain boundary pinning, created by the densification procedures and the mechanical deformation applied.

Keywords: iron-based superconductors; critical currents; flux pinning; microstructure

1. Introduction

The discovery of iron-based superconductors (IBS) did not only initiate new research on the origin of superconductivity but also boosted new research on materials for applications [1]. Here, mainly three IBS compounds are important, the 1111-type (ROFeAs with R = denoting a rare earth element) materials, the 122-compounds (AFe_2As_2 with A = alkaline earth) and the very simple iron-chalcogenides of the 11-type (e.g., FeSe). From all these compounds, thin films, as well as tapes and bulk samples, were already produced [1–3].

The compound $Ba_{0.6}K_{0.4}Fe_2As_2$ of the 122-family (hereafter abbreviated as 122) offers certain advantages as compared to cuprates, MgB_2 and other IBS materials. The superconducting transition temperature, T_c, is higher than that of FeSe and it is less anisotropic than the 1111 compounds. Regarding possible applications, the 122 compounds can operate at higher magnetic fields than MgB_2 [2], are less prone to flux jumps, are robust to impurity doping [3], and the misalignment of the grains can be higher than in cuprates when producing wires or bulks. These advantages were already manifested in several efforts to produce 122-based wires/tapes (see, e.g., the recent review of Yao et al. [4]), and trapped fields above 1 T were already recorded in bulk, HIP (hot isostatic pressed)-processed 122 samples [5]. On the other hand, the 122 material demonstrates several unique properties, e.g., it undergoes an isostructural transition under pressure [6–9], from which important information on the nature of superconductivity can be obtained.

In a previous review article, the flux pinning properties ($F_p = j_c \times B$, with j_c denoting the critical current density, and $B = \mu_0 H_a$), and especially, the scaling properties of the flux pinning forces of various IBS materials were discussed [10]. The 122 compounds analyzed showed a tendency towards dominant flux pinning of the δl-type, and a contribution of δT_c-pinning could not be excluded (that is, the peak position of the scaling diagrams ranged between 0.25 and 0.5). For the tapes, a further improvement of densification and deformation is an essential issue creating continuous current flow via many grain boundaries (GBs), and also, flux pinning may be provided by GBs as in the case of MgB_2. This was discussed in detail in several publications [11–14]. The recent advances in j_c of tapes [4] and bulk materials [5,15,16] are, therefore, due to these improved sample preparation routes.

Based on these developments, in the present contribution, we discuss the flux pinning properties measured on reacted-and-pressed 122 powder samples, and analyze the microstructural properties of the 122 grains by means of electron backscatter diffraction (EBSD). This combination of detailed microstructural analysis and magnetic data has already been proven to be very effective in previous publications [17,18]. Furthermore, we compare our data with recent data of 122 bulks and tapes.

2. Experimental Procedures

The $Ba_{0.6}K_{0.4}Fe_2As_2$ (122) powders were prepared using stoichiometric mixtures of the elements (purities > 99.9%) in alumina crucibles, which were sealed in silica tubes in an argon atmosphere. To prevent the evaporation of potassium during heat treatment, alumina inlays were employed. The mixtures were heated up to 600 °C (heating rate 50 °C/h). This temperature was maintained for 15 h. The reaction products were ground again and homogenized, sealed in a silica tube, and subjected to heat treatment at 650 °C for 15 h. Then, the mixtures were ground and homogenized and cold-pressed into pellets. Finally, the samples were sintered at 750 °C (20 h). The samples were furnace-cooled down to room temperature. The potassium evaporation caused only small fractions of the impurity phase FeAs (<3.5%), which was detected by X-ray analysis (Cu-$K_{\alpha 1}$-radiation). The lattice parameters, the Ba/K atomic ratio, and the crystal purity of the 122 powders were already reported in [19–21]. Finally, to obtain the pellets studied here, the powders were again cold-pressed into pellets with 5 mm diameter. The X-ray data and the description of the preparation of ex-situ 122 tapes manufactured from the same powders are given in [21].

For the microstructural analysis, the sample surfaces were mechanically polished in dry conditions using various 3M imperial lapping sheets down to 0.25 μm roughness. A final polishing step using a colloidal silica solution (Struers water-free OP-AA solution) with a grain size of 25 nm was carried out. Details of the polishing procedure were already described in detail [22,23]. For the EBSD analysis, the surfaces of the samples were further subjected to low-angle argon ion-polishing (5 keV, 5 min) in order to improve the image quality of the Kikuchi patterns further. This procedure was described already for the investigation of multiferroic samples in [24,25].

The EBSD analysis was performed using a JEOL 7000F SEM microscope equipped with an orientation imaging analysis unit (EDAX Inc., OIM Analysis™) [26]. The Kikuchi patterns were generated at an acceleration voltage of 15 kV, and were recorded by means of a DigiView camera system. Automated EBSD mappings were carried out using step sizes down to 50 nm. The working distance was set to 15 mm.

Magnetic measurements were performed using a Quantum Design MPMS3 SQUID system with ±7 T applied field, and Quantum Design physical property measurement system (PPMS) systems equipped either with an extraction magnetometer or with a vibrating sample magnetometer option (±7 T and ±9 T applied magnetic field). In all cases, the field was applied perpendicular to the sample surface. The field sweep rate was always 0.36 T/min. From the magnetization data, the critical current densities, j_c, were evaluated using the extended Bean approach for rectangular samples [27] and the extensions described in [28–30]. The irreversibility lines were determined using a criterion of 100 A/cm^2 from the obtained measurements, except the literature data [16], where extrapolation from the $j_c(H)$ graphs was employed [31].

3. Results and Discussion

Figure 1 presents a DC susceptibility measurement of the reacted-and-pressed 122 powder sample in a zero-field cooled (ZFC) condition with an applied field of 10 mT. The transition temperature, T_c, was found to be 38.5 K and a relatively sharp superconducting transition was observed.

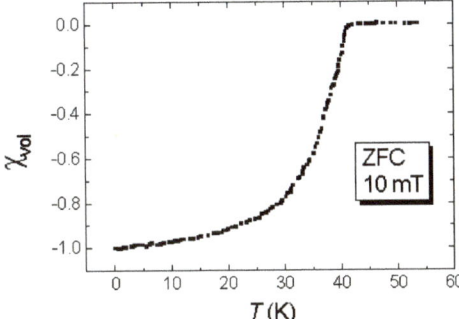

Figure 1. DC volume susceptibility, χ_{vol}, as function of temperature measured in a zero-field cooled (ZFC) condition (applied field 10 mT).

Figure 2 gives a double logarithmic plot of the critical current density, j_c, as a function of the applied field, H_a, in the temperature range 25 K to 35 K. Note that this current density is a true intra-grain current density only, as in the reacted-and-pressed powder sample no contribution of a transport current across the grain boundaries is expected. The flat part at low fields corresponds to single vortex pinning, which is then followed by a narrow regime with an intermediate curvature, and then a steep decrease towards the irreversibility is seen which is dominated by flux-flow. An extrapolation of this plot towards the 100 A/cm²-line gives information on the irreversibility fields used for the pinning force scaling below. The nominal values of j_c obtained for the reacted-and-pressed powder sample of 10^4 A/cm² at 25 K and self-field are comparable to those of polycrystalline, bulk FeSe [32,33], but are clearly smaller than those of ex-situ powder-in-tube tapes produced using the same 122 powder. Here, a current density of 3×10^4 A/cm² was obtained at 4.2 K, using transport measurement on a 10 cm-long piece of tape with a filling factor of 30% [21].

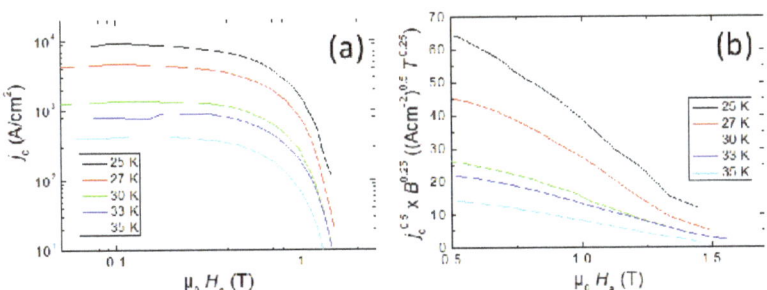

Figure 2. (a) double logarithmic plot of j_c as a function of the applied field, $\mu_0 H_a$; (b) Kramer plot of the same data set.

To obtain more information about the flux pinning properties, Kramer plots, i.e., $j_c^{0.5} \times B^{0.25}$ versus the applied field, $B = \mu_0 H_a$, are employed [34]. The advantage here is that the information obtained is free of uncertainties such as the determination of H_{irr}. This is especially important when comparing data of different origins, i.e., unknown measurement parameters such as the field sweep rate. In Figure 2b, the Kramer plot for the same dataset is presented. At applied fields above 0.6 T, all curves are found to be quasi linear. An extrapolation of these curves yields information about the upper critical field, H_{c2}. It is obvious that H_{c2} obtained in this way is always larger than the irreversibility fields, as shown below.

For comparison of our data with those of a fully-reacted pnictide sample of the same composition, Figure 3a shows a double-log plot of the j_c-data by Weiss et al. [16] plotted in a similar manner to Figure 2. It is clearly visible that the critical current density obtained here is much higher, and the field dependence is very different, indicating high irreversibility fields. The magneto-optic images of flux penetration presented in [16] demonstrate that a homogeneous flux front penetrates the sample when applying external magnetic fields. This manifests as well-developed grain coupling, so the intergranular critical current density in this sample is high, and is clearly the predominant contribution to the overall j_c measured with the magnetization data. Furthermore, the much higher H_{irr} data reveal that the flux pinning created by the mechanical deformation is predominant even up to high applied fields. Figure 3b gives the corresponding Kramer plot. Here, linear sections are also obtained. Note that the low-temperature data show an increase with increasing field, which indicates that the applied field range is still lower than the maximum pinning force ($F_p = j_c \times B$). The 122 sample of Weiss et al. was treated using hot-isostatic pressing (HIP), resulting in a dense sample microstructure with only a minor number of voids present [16]. The comparison of the two data sets indicates that the improved j_c and H_{irr} values are due to densification and deformation during the HIP process.

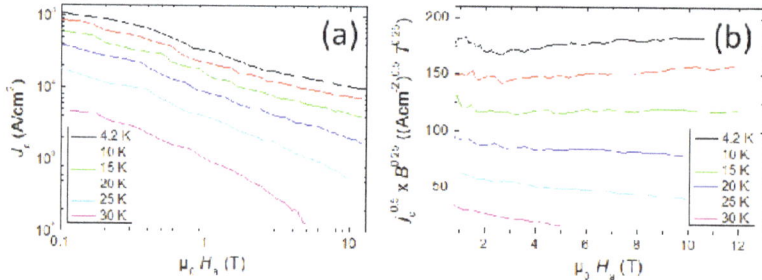

Figure 3. (a) double logarithmic plot of j_c as a function of the applied field, $\mu_0 H_a$; (b) Corresponding Kramer plot of the same data set.

In Figure 4a, the pinning force scaling following the approach of Dew-Hughes (DH) [35] of the reacted-and-pressed powder sample is shown. The obtained scaling is well developed, yielding a pinning function $Ah^p \times (1-h)^q$ with $p = 1.5$ and $q = 3.05$. The parameter A is not a real fit parameter, but defined via the condition $F_p/F_{p,\max}(h_0) = 1$ [36]. Following the review on pinning force scaling of various IBS materials [10], this scaling is very similar to that of different single crystals and thin film samples of the 122 compounds. The resulting peak position, h_0, is obtained at 0.33, which indicates a dominant flux pinning provided by small normal-conducting inclusions.

Figure 4b illustrates the pinning force scaling of the data from Weiss et al. [16]. The irreversibility fields used for this scaling were obtained from extrapolations of the double-log plot as shown in Figure 3a, and finally, the data were adjusted by the scaling itself. For this procedure, we could only employ the data of 20, 25 and 30 K. For all other data, the measured field range is simply too small. Figure 4b demonstrates that it is possible to force the data to a scaling behavior, but it is difficult to obtain a proper fit to the data. The peak position, h_0, is only reached by the data taken at 25 K and 30 K. Here, it is clear that the scaling parameter p must be smaller than 1 to model the shape of the low-field part seen here, and the results of this attempt are depicted in Figure 4b using the red fit (1) with $p = 0.6$, $q = 1.62$ and the green fit (2) with $p = 0.3$, $q = 0.81$. Furthermore, the high-field tail of the 30 K data clearly deviates from any DH pinning function. For both fits, the peak is found at $h_0 = 0.27$, which is smaller than that of the reacted-and-pressed powder samples, but with a strong low-field side which does not reveal the typical curvature of the DH functions. The scaling parameters obtained here do not fit the DH model exactly, as the parameter q describes the type of pinning, and hence, takes only the values of 1 and 2 for superconducting and normal-conducting pinning sites, respectively. In Table 1, the values of q for the tapes are close to two, but deviations are seen for the reacted-and-pressed powder sample, as well as for fits (1) and (2) to the HIP-processed bulk sample. Values for $q > 2$ were often observed in 123-type cuprate superconductors [37,38] and also, in IBS materials [10], but a $q < 0.5$ (fit (2)) is outstanding, which could indicate a different pinning source; the DH function with $p = 0.5$ and $q = 1$ would describe magnetic volume pinning [35]. The parameter p can range between 0 and 2 and gives information about the extension of the flux pinning sites. A value of $p = 0.5$ corresponds to surface pinning (=GBs, 2D-pinning by extended defects), and $p = 0$ describes volume pinning, which would indicate that fit (2) describes pinning by GB clusters.

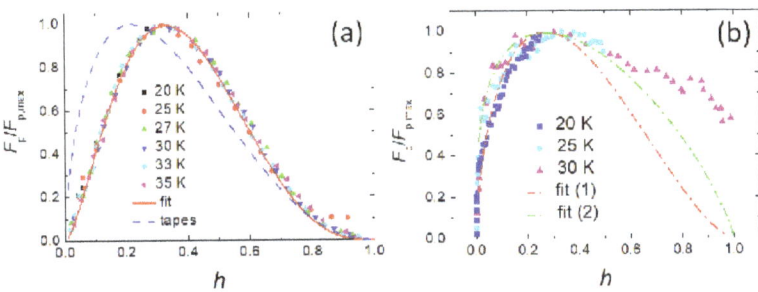

Figure 4. Pinning force scaling, $F_p/F_{p,\max}$ versus the reduced field, $h = H_a/H_{\mathrm{irr}}$, for the reacted-and-pressed 122 powder sample (**a**) and the HIP-processed 122 sample by Weiss et al. [16] (**b**). In (**a**), the best fit to the data is shown by a red line (—). For comparison, the fit curve obtained on 122 tapes [3] is shown with a hatched blue line (– –). In (**b**), the two fits (1), (2) are attempts to fit the data by the DH model (for the parameters employed, see the main text).

Table 1. Fitting parameters and peak positions of the DH scaling of various 122 sample types. For the references, see the main text.

Sample	p	q	h_0	Remarks
reacted-and-pressed	1.5	3.05	0.33	good scaling, similar to 122 crystals
HIP-processed bulk				
fit (1)	0.6	1.62	0.27	fit close to the low-field data at all T
fit (2)	0.3	0.81	0.27	steeper fit, 20 K data deviate clearly
tapes				
Huang	0.64	2.3	0.22	good scaling, 20 K $\leq T \leq$ 32 K
Shabbir (0 GPa)	0.65	1.95	0.25	$T = 24$ K
Shabbir (0.7 GPa)	0.8	2	0.29	$T = 24$ K, hydrostatic pressure applied

Furthermore, Figure 4a gives the result of the pinning force scaling of 122-tapes as determined by Huang et al. [39]. These authors obtained a good scaling in the range 25 K $< T <$ 35 K with $p = 0.64$ and $q = 2.3$, yielding a peak position $h_0 = 0.22$. This clearly indicates that the predominant flux pinning is provided by the grain boundaries. A similar result was obtained by Shabbir et al. [40], where $p = 0.65$ and $q = 1.95$ (0 GPa) and $p = 0.8$ / $q = 2$ (0.7 GPa hydrostatic pressure) was obtained. These authors studied the application of hydrostatic pressure to densify the superconducting material. The change in the fitting parameters of the pinning force scaling (however, only measured at $T = 24$ K) reveals that the pressure induced more point defects into the material, which improve the flux pinning properties even further, thereby, a record high j_c was obtained at 4.2 K.

Figure 5 presents the irreversibility fields, H_{irr}, obtained for the reacted-and-pressed powder sample (red bullets), the data extracted from the measurements of Weiss et al. (black squares) and the data from Huang et al. [39] (blue line) for 122 tapes. The comparison here is straightforward as all samples exhibit a similar value of T_c, in contrast to the comparison presented in [33]. The reacted-and-pressed 122 sample shows only small irreversibility fields in comparison with the other sample types (bulks, tapes). At temperatures close to T_c, the tape-data and those of Weiss are similar to each other, and at lower temperatures, the data of Weiss are much higher, which may also be due to the different measurement techniques (magnetic vs. transport). When fitting the data using the relation $H_{\text{irr}}(T) = H_0(1 - T/T_c)^n$ with $n = 3$, common for many 123-high-T_c superconducting materials [41], we obtain a fit parameter $H_0 = 180$ T (fit (1)). To fit the data of Weiss et al., we can use also $H_0 = 180$ T, but n is changed to 1.65 (fit (2)). This behavior points again to the fact that the origin of flux pinning of the reacted-and-pressed 122 powder is similar to that in cuprate superconductors, especially YBa$_2$Cu$_3$O$_7$ (YBCO), where a peak position of $h_0 = 0.33$ is also obtained. Thus, a weak collective pinning by point defects is predominant. However, in contrast to the YBCO compounds, the j_c and H_{irr} values in the reacted-and-pressed powder sample are small, whereas both values are high in YBCO, even for polycrystalline samples. In the case of the 122 samples for applications that have underwent densification/mechanical treatments, a different origin of the flux pinning is revealed as indicated by the shift of the peak position, h_0, of the pinning force scaling diagram towards smaller values close to 0.2. This observation points to the strong role of the mechanical deformation [42] and densification processes to develop a strong flux pinning in the 122 materials.

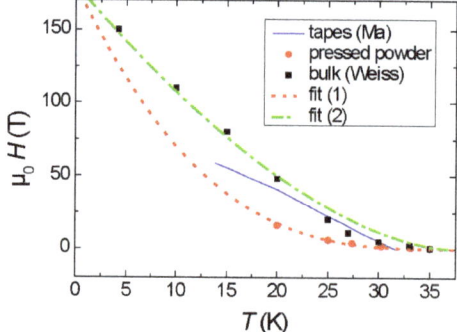

Figure 5. Irreversibility fields, H_{irr}, for the reacted-and-pressed powder sample (●), tapes indicated by a blue line (—) [39], and the HIP-treated bulk sample (■) of Weiss et al. [16]. Details of the fits (1) and (2) are described in the main text.

The critical current densities measured in the reacted-and-pressed powder sample are truly intragrain current densities, as there is no current flow across the GBs. From the data presented here, we can conclude that the intragrain flux pinning is provided by a weak δl-pinning (small normal-conducting inclusions or defects), which also determines the small values of H_{irr} and H_{c2}. In contrast, the strong pinning and the large irreversibility fields found in the deformed (densified) 122 samples is provided by GB pinning, which neither exists in the reacted-and-pressed 122 powder samples nor in the 122 single crystals investigated in the literature. Furthermore, this GB pinning is also responsible for the high irreversibility fields of the 122 tapes and bulks. These observations are in contrast to the cuprate superconductors, and here, especially YBCO, where the intragranular current density of the grains is always high, and also dominates H_{irr}.

Now, we turn to the microstructural analysis of the reacted-and-pressed powder sample by means of EBSD. This provides the possibility to compare the data with those obtained on 122 tapes already published in the literature [39,40].

Figure 6 presents the Kikuchi patterns of the 122 compound and the EBSD indexation. The resulting Euler angles are $\phi_1 = 316.1°$, $\Phi = 0.9°$ and $\phi_2 = 114.3°$. The image quality (IQ) is high at ~6000, and the confidence index for the indexation is 1 (i.e., exact indexation).

In Figure 7a–g, EBSD mappings and dataplots are presented. Figure 7a is an image quality map, which resembles a backscattered electron image, but in EBSD conditions (70° tilt), of the sample. Here, we see that there are many large grains, but also a high number of small grains. There is a large amount of voids and cracks in the investigated sample area, typical for such a reacted-and-pressed powder sample. Figure 7b presents an orientation mapping in the direction normal to the sample surface (i.e., in [001]-direction or ND). The color code for this map is given in the stereographic triangle below. The result of the grain orientation mapping is also presented in the corresponding pole figure (c), showing several distinct maxima. The map (d) is a grain size map with grain sizes ranging from 0.1 μm (blue) to 4 μm (red). This is further illustrated in the corresponding grain size plot (f). The grain sizes range between 110 nm and 3.2 μm. The map (e) gives the distribution of the grain shape aspect ratio, γ_{ar}. To obtain this information, the 122 grains in the orientation map (b) are encircled by ellipses by the EBSD software. The grain shape aspect ratio, γ_{ar}, is defined as the length of the minor axis divided by the length of the major axis. The value of γ_{ar} ranges between 0 and 1, with higher values indicating similar lengths of the axes [26]. The color code for this map is indicated on top of it. The plot (g) gives the statistics of the EBSD-determined misorientation angles (number fraction vs. misorientation angle). Here, a large number of small-angle misorientations is obtained, and also a maximum at ~37°. Finally, the results of the grain shape aspect ratio analysis are presented in plot (h). The majority of the 122 grains are only slightly elliptic; grains with an elongated major axis γ_{ar} between 0.1 and 0.4 are quite rare in the investigated area. Comparing our results of the EBSD-analysis of the reacted-and-pressed powder sample to those of 122-tapes published in the literature [39,40], we see that the 122 grains in the tapes show a more elliptic grain aspect, which is a consequence of the deformation treatment applied. This is a remarkable result of the EBSD investigation, and is important for the processing of future 122 samples for applications. The development of the specific microstructure in the tape samples and that of HIP-processed bulks will, therefore, require more detailed investigations in the future to further improve the performance of the 122 superconductor samples.

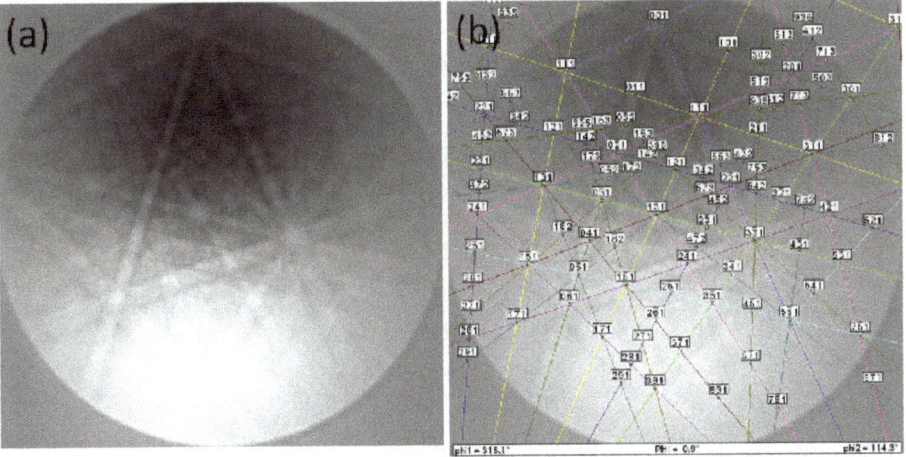

Figure 6. (a) Kikuchi pattern of the 122 compound and (b) its indexation (Image quality ~6000, CI = 1). The resulting Euler angles are $\phi_1 = 316.1°$, $\Phi = 0.9°$ and $\phi_2 = 114.3°$.

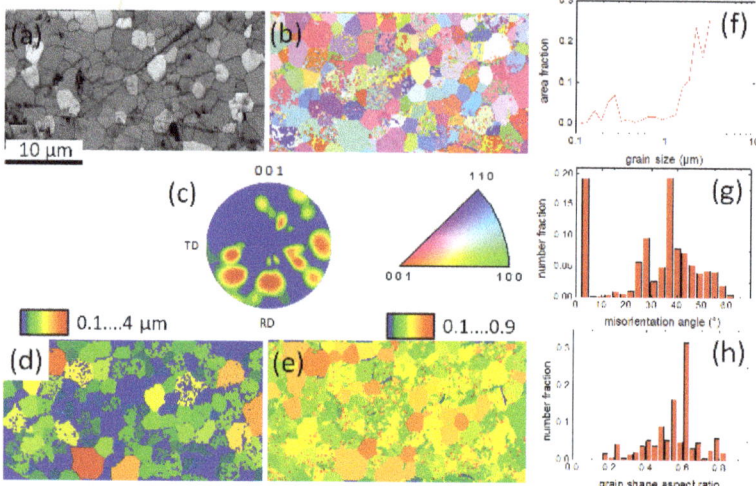

Figure 7. Electron backscatter diffraction (EBSD) analysis of the 122 reacted-and-pressed powder sample. (**a**) is an image quality (IQ) map; (**b**) gives the orientation mapping in the direction normal to the sample surface. The color code for this map is given in the stereographic triangle below the map; (**c**) presents the corresponding pole figure in the [001]-direction (i.e., normal to the sample surface, or ND); (**d**) is a grain size mapping, and (**e**) is a map of the grain size aspect ratio. The color codes for the maps (**d**,**e**) are indicated at the top of each map. The plots (**f**–**h**) illustrate details of the sample microstructure. (**f**) gives the statistics of the EBSD-determined grain size, (**g**) the distribution of the misorientation angles, and (**h**) the statistics of the grain shape aspect ratio. For details, see the main text.

The results obtained on the reacted-and-pressed powder sample corroborate that the mechanical deformation and densification processes (e.g., HIP, wire drawing and rolling) play an essential role in order to achieve the desired flux pinning properties, and not only the improvement of texture, such as in cuprate superconductors, where the intragrain critical current density (j_c(intra)) always plays the predominant role. Both the critical current densities (flux pinning forces) and the irreversibility fields develop their strength in the 122 samples upon occurrence of mechanical deformation processes, whereas the reacted-and-pressed 122 powder sample is found to behave much more similarly to single crystals of the same compound [10]. The comparison of the EBSD results on the reacted-and-pressed powder samples with that obtained on 122 tapes reveals changes in the grain shape that were induced by the processing. Therefore, to achieve even higher critical current densities (flux pinning forces), more efforts to study the microstructural changes in the mechanical deformation processes of the 122-grains are required.

4. Conclusions

To conclude, we have presented magnetic and microstructural data of a reacted-and-pressed 122 powder sample and performed a comparison of our results with the literature data of a HIP-processed 122 bulk sample and 122 tapes. The obtained critical current densities (which are intragrain only) and the irreversibility fields of the reacted-and-pressed powder sample are much lower as compared to the bulks and tapes. The pinning force scaling of the reacted-and-pressed powder sample is well developed, with a dominant pinning provided by normal-conducting small inclusions (peak position at 0.33), similar to 122 single crystals. This is in contrast to the results obtained on HIP-processed 122 bulks and 122 tapes, where much lower peak positions in the pinning force scaling diagrams are obtained (0.27 and 0.22). This points to a dominant flux pinning provided by grain boundaries, so that

j_c(inter) > j_c(intra). Therefore, the mechanical deformation and densification processes induced in manufacturing samples for applications are essential to achieve high critical currents and high irreversibility fields in the 122 samples.

Author Contributions: Conceptualization, M.R.K.; Formal Analysis, A.K.-V. and M.R.K.; Investigation, A.K.-V., M.R.K. and J.S.; Supervision, M.M.; Writing—Original Draft Preparation, M.R.K.; Writing—Review and Editing, A.K.-V. and M.R.K.

Funding: This work is part of the SUPERFOAM international project funded by ANR and DFG under the references ANR-17-CE05-0030 and DFG-ANR Ko2323-10, respectively.

Acknowledgments: We thank J. Wiesenmayer and D. Johrendt (LMU, München, Germany) for providing us with the polycrystalline $Ba_{0.6}K_{0.4}Fe_2As_2$ sample, C. Chang (IJL Nancy, France) for help with some of the magnetic measurements and K. Berger, Q. Nouailhetas, B. Douine (GREEN, Nancy, France) for valuable discussions.

Conflicts of Interest: The authors declare no conflict of interest.

References

1. Hosono, H.; Yamamoto, A.; Hiramatsu, H.; Ma, Y. Recent advances in iron-based superconductors toward applications. *Mater. Today* **2018**, *21*, 278–302. [CrossRef]
2. Gurevich, A. Iron-based superconductors at high magnetic fields. *Rep. Prog. Phys.* **2011**, *74*, 124501. [CrossRef]
3. Ma, Y. Progress in wire fabrication of iron-based superconductors. *Supercond. Sci. Technol.* **2012**, *25*, 113001. [CrossRef]
4. Yao, C.; Ma, Y. Recent breakthrough development in iron-based superconducting wires for practical applications. *Supercond. Sci. Technol.* **2019**, *32*, 023002. [CrossRef]
5. Weiss, J.D.; Yamamoto, A.; Polyanskii, A.A.; Richardson, R.B.; Larbalestier, D.C.; Hellstrom, E.E. Demonstration of an iron-pnictide bulk superconducting magnet capable of trapping over 1 T. *Supercond. Sci. Technol.* **2015**, *28*, 112001. [CrossRef]
6. Uhoya, W.; Stemshorn, A.; Tsoi, G.; Vohra, Y.K.; Sefat, A.S.; Sales, B.C.; Hope, K.M.; Weir, S.T. Collapsed tetragonal phase and superconductivity of $BaFe_2As_2$ under high pressure. *Phys. Rev. B* **2010**, *82*, 144118. [CrossRef]
7. Mittal, R.; Mishra, S.K.; Chaplot, S.L.; Ovsyannikov, S.V.; Greenberg, E.; Trots, D.M.; Dubrovinsky, L.; Su, Y.; Brueckel, T.; Matsuishi, S.; et al. Ambient- and low-temperature synchrotron x-ray diffraction study of $BaFe_2As_2$ and $CaFe_2As_2$ at high pressures up to 56 GPa. *Phys. Rev. B* **2011**, *83*, 054503. [CrossRef]
8. Nakajima, Y.; Wang, R.; Metz, T.; Wang, X.; Wang, L.; Cynn, H.; Weir, S.T.; Jeffries, J.R.; Paglione, J. High-temperature superconductivity stabilized by electron-hole interband coupling in collapsed tetragonal phase of KFe_2As_2 under high pressure. *Phys. Rev. B* **2015**, *91*, 060508. [CrossRef]
9. Ptok, A.; Sternik, M.; Kapcia, K.J.; Piekarz, P. Structural, electronic, and dynamical properties of the tetragonal and collapsed tetragonal phases of KFe_2As_2. *Phys. Rev. B* **2019**, *99*, 134103. [CrossRef]
10. Koblischka, M.R.; Muralidhar, M. Pinning force scaling analysis of Fe-based high-T_c superconductors. *Int. J. Mod. Phys. B* **2016**, *30*, 1630017. [CrossRef]
11. Katase, T.; Ishimaru, Y.; Tsukamoto, A.; Hiramatsu, H.; Kamiya, T.; Tanabe, K.; Hosono, H. Advantageous grain boundaries in iron pnictide superconductors. *Nat. Commun.* **2011**, *2*, 409. [CrossRef] [PubMed]
12. Kametani, F.; Li, P.; Abraimov, D.; Polyanskii, A.A.; Yamamoto, A.; Jiang, J.; Hellstrom, E.E.; Gurevich, A.; Larbalestier, D.C.; Ren, Z.A.; et al. Intergrain current flow in a randomly oriented polycrystalline $SmFeAsO_{0.85}$ oxypnictide. *Appl. Phys. Lett.* **2009**, *95*, 142502. [CrossRef]
13. Hecher, J.; Baumgartner, T.; Weiss, J.D.; Tarantini, C.; Yamamoto, A.; Jiang, J.; Hellstrom, E.E.; Larbalestier, D.C.; Eisterer, M. Small grains: A key to high-field applications of granular Ba-122 superconductors? *Supercond. Sci. Technol.* **2016**, *29*, 025004. [CrossRef]
14. Iida, K.; Hänisch, J.; Tarantini, C. Fe-based superconducting thin films on metallic substrates: Growth, characteristics, and relevant properties. *Appl. Phys. Rev.* **2018**, *5*, 031304. [CrossRef]
15. Weiss, J.D.; Tarantini, C.; Jiang, J.; Kametani, F.; Polyanskii, A.A.; Larbalestier, D.C.; Hellstrom, E.E. High intergrain critical current density in fine-grain $(Ba_{0.6}K_{0.4})Fe_2As_2$ wires and bulks. *Nat. Mater.* **2012**, *11*, 682–685. [CrossRef]

16. Weiss, J.D.; Jiang, J.; Polyanskii, A.A.; Hellstrom, E.E. Mechanochemical synthesis of pnictide compounds and superconducting Ba$_{0.6}$K$_{0.4}$Fe$_2$As$_2$ bulks with high critical current density. *Supercond. Sci. Technol.* **2013**, *26*, 074003. [CrossRef]
17. Koblischka-Veneva, A.; Koblischka, M.R.; Qu, T.; Han, Z.; Mücklich, F. Texture analysis of monofilamentary, Ag-sheathed (Pb,Bi)$_2$Sr$_2$Ca$_2$Cu$_3$O$_x$ tapes by electron backscatter diffraction (EBSD). *Physica C* **2008**, *468*, 174–182. [CrossRef]
18. Koblischka-Veneva, A.; Koblischka, M.R.; Schmauch, J.; Inoue, K.; Muralidhar, M.; Berger, K.; Noudem, J. EBSD analysis of MgB$_2$ bulk superconductors. *Supercond. Sci. Technol.* **2016**, *29*, 044007. [CrossRef]
19. Wiesenmayer, E.; Luetkens, H.; Pascua, G.; Khasanov, R.; Amato, A.; Potts, H.; Banusch, B.; Klauss, H.-H.; Johrendt, D. Microscopic Coexistence of Superconductivity and Magnetism in Ba$_{1-x}$K$_x$Fe$_2$As$_2$. *Phys. Rev. Lett.* **2011**, *107*, 237001. [CrossRef]
20. Wiesenmayer, J.E. Dissertation LMU München, Germany. Available online: https://edoc.ub.uni-muenchen.de/18509/1/Wiesenmayer_Josef_Erwin.pdf (accessed on 20 May 2019).
21. Malagoli, A.; Wiesenmayer, E.; Marchner, S.; Johrendt, D.; Genovese, A.; Putti, M. Role of heat and mechanical treatments in the fabrication of superconducting Ba$_{0.6}$K$_{0.4}$Fe$_2$As$_2$ ex situ powder-in-tube tapes. *Supercond. Sci. Technol.* **2015**, *28*, 095015. [CrossRef]
22. Koblischka, M.R.; Koblischka-Veneva, A. Applications of the electron backscatter diffraction technique to ceramic materials. *Phase Transit.* **2013**, *86*, 651–660. [CrossRef]
23. Koblischka, M.R.; Koblischka-Veneva, A. Advanced Characterization of Multiferroic Materials by Scanning Probe Methods and Scanning Electron Microscopy. In *Nanoscale Ferroelectrics and Multiferroics*; Alguero, M., Gregg, J.M., Mitoseriu, L., Eds.; Wiley and Sons Ltd.: Chichester, UK, 2016; pp. 400–434.
24. Koblischka-Veneva, A.; Koblischka, M.R.; Schmauch, J.; Chen, Y.; Harris, V.G. EBSD analysis of the microtexture of Ba-hexaferrite samples. *J. Phys. Conf. Ser.* **2010**, *200*, 082014. [CrossRef]
25. Koblischka-Veneva, A.; Koblischka, M.R.; Chen, Y.; Harris, V.G. Analysis of Grain Shape and Orientation in BaFe$_{12}$O$_{19}$-Ferrites Using Electron Backscatter Diffraction (EBSD). *IEEE Trans. Magn.* **2009**, *45*, 4219–4222. [CrossRef]
26. *Orientation Imaging Microscopy (OIM Analysis™)*; Software Version V8.1, User Manual; EDAX Inc.: Draper, UT, USA, 2018.
27. Wiesinger, H.P.; Sauerzopf, F.M.; Weber, H.W. On the calculation of J_c from magnetization measurements on superconductors. *Physica C* **1992**, *203*, 121–128. [CrossRef]
28. Gokhfeld, D.M.; Balaev, D.A.; Petrov, M.I.; Popkov, S.I.; Shaykhutdinov, K.A.; Val'kov, V.V. Magnetization asymmetry of type-II superconductors in high magnetic fields. *J. Appl. Phys.* **2011**, *109*, 033904. [CrossRef]
29. Gokhfeld, D.M. Secondary Peak on Asymmetric Magnetization Loop of Type-II Superconductors. *J. Supercond. Novel Magn.* **2013**, *26*, 281–283. [CrossRef]
30. Gokhfeld, D.M. The circulation radius and critical current density in type-II superconductors. *Tech. Phys. Lett.* **2019**, *45*, 1–3. [CrossRef]
31. Sun, Y.; Pyon, S.; Tamegai, T.; Kobayashi, R.; Watashige, T.; Kasahara, S.; Matsuda, Y.; Shibauchi, T.; Kitamura, H. Enhancement of critical current density and mechanism of vortex pinning in H$^+$-irradiated FeSe single crystal. *Appl. Phys. Exp.* **2015**, *8*, 113102. [CrossRef]
32. Karwoth, T.; Furutani, K.; Koblischka, M.R.; Zeng, X.L.; Wiederhold, A.; Muralidhar, M.; Murakami, M.; Hartmann, U. Electrotransport and magnetic measurements on bulk FeSe superconductors. *J. Phys. Conf. Ser.* **2018**, *1054*, 012018. [CrossRef]
33. Koblischka-Veneva, A.; Koblischka, M.R.; Berger, K.; Nouailhetas, Q.; Douine, B.; Muralidhar, M.; Murakami, M. Comparison of Temperature and Field Dependencies of the Critical Current Densities of Bulk YBCO, MgB$_2$, and Iron-Based Superconductors. *IEEE Trans. Appl. Supercond.* **2019**, *29*, 6801805. [CrossRef]
34. Kramer, E.J. Scaling laws for flux pinning in hard superconductors. *J. Appl. Phys.* **1973**, *44*, 1360–1370. [CrossRef]
35. Dew-Hughes, D. Flux pinning mechanisms in type-II superconductors. *Philos. Mag.* **1974**, *30*, 293–305. [CrossRef]
36. Koblischka, M.R. Pinning in bulk high-T_c superconductors. *Inst. Phys. Conf. Ser.* **1997**, *58*, 1141–1144.
37. Koblischka, M.R.; van Dalen, A.J.; Higuchi, T.; Yoo, S.I.; Murakami, M. Analysis of pinning in NdBa$_2$Cu$_3$O$_{7-\delta}$ superconductors. *Phys. Rev. B* **1998**, *58*, 2863–2867. [CrossRef]

38. Koblischka, M.R.; Muralidhar, M. Pinning force scaling and its analysis in the LRE-123 ternary compounds. *Physica C* **2014**, *496*, 23–27. [CrossRef]
39. Huang, H.; Yao, C.; Dong, C.H.; Zhang, X.P.; Wang, D.L.; Cheng, Z.; Li, J.Q.; Awaji, S.; Wen, H.H.; Ma, Y.W. High transport current superconductivity in powder-in-tube $Ba_{0.6}K_{0.4}Fe_2As_2$ tapes at 27 T. *Supercond. Sci. Technol.* **2018**, *31*, 015017. [CrossRef]
40. Shabbir, B.; Huang, H.; Yao, C.; Ma, Y.W.; Dou, S.X.; Johansen, T.H.; Hosono, H.; Wang, X.L. Evidence for superior current carrying capability of iron pnictide tapes under hydrostatic pressure. *Phys. Rev. Mater.* **2017**, *1*, 044805. [CrossRef]
41. Altin, E.; Gokhfeld, D.M.; Demirel, S.; Oz, E.; Kurt, F.; Altin, S.; Yakinci, M.E. Vortex pinning and magnetic peak effect in $Eu(Eu,Ba)_{2.125}Cu_3O_x$. *J. Mater. Sci. Mater. Electron.* **2014**, *25*, 1466–1473. [CrossRef]
42. Han, Z.; Skov-Hansen, P.; Freltoft, T. The mechanical deformation of superconducting BiSrCaCuO/Ag composites. *Supercond. Sci. Technol.* **1997**, *10*, 371–387. [CrossRef]

© 2019 by the authors. Licensee MDPI, Basel, Switzerland. This article is an open access article distributed under the terms and conditions of the Creative Commons Attribution (CC BY) license (http://creativecommons.org/licenses/by/4.0/).

Article

Study on Quenching Characteristics and Resistance Equivalent Estimation Method of Second-Generation High Temperature Superconducting Tape under Different Overcurrent

Siyuan Liang [1], Li Ren [1,*], Tao Ma [2], Ying Xu [1], Yuejin Tang [1], Xiangyu Tan [1], Zheng Li [1], Guilun Chen [1], Sinian Yan [1], Zhiwei Cao [1], Jing Shi [1], Leishi Xiao [3] and Meng Song [3]

1. State Key Laboratory of Advanced Electromagnetic Engineering and Technology, School of Electrical and Electronic Engineering, Huazhong University of Science and Technology, Wuhan 430074, China
2. School of Electrical Engineering, Beijing Jiaotong University, Beijing 100044, China
3. Electric Power Research Institute of Guangdong Power Grid Corporation, Guangzhou 510080, China
* Correspondence: renli@mail.hust.edu.cn; Tel.: +86-027-87544755

Received: 6 June 2019; Accepted: 24 July 2019; Published: 25 July 2019

Abstract: In this paper, through AC and DC overcurrent tests on second generation high temperature superconducting tape (2G HTS tape), we respectively summarize the typical types of quenching resistance and corresponding quenching degree, in which there are three types under AC overcurrent and two types under DC overcurrent. According to experimental results, a rule was found that, when 2G HTS tape quenches to normal state, the relationship between quenching resistance and joule heat generated from 2G HTS tape presents a fixed trend line, and the influence of liquid nitrogen can be ignored. Then, the characteristics and rules of quenching resistance found in experiments are well explained and confirmed by a detailed 3D finite element model of 2G HTS tape including electromagnetic field and thermal field. Finally, based on above works, our group proposes a new equivalent method to estimate the quenching resistance, where the results of AC and DC overcurrent experiments can be equivalent to each other within a certain range. Compared with FEM, the method has the following advantages: (i) The method is simple and easy to implement. (ii) This method combines precision and computational efficiency. (iii) With superconducting tape quenching to normal state, this method presents a good consistency with experimental results.

Keywords: superconducting tape; quench; R-SFCL; AC and DC overcurrent; experiment; finite element method (FEM); numerical modeling

---

1. Introduction

With the discovery of the cuprate-based high temperature superconductors, first generation high temperature superconducting (1G HTS) tapes represented by BSCCO (Bismuth Strontium Calcium Copper Oxide) Ag-sheathed conductors and second-generation high temperature superconducting (2G HTS) tapes represented by YBCO (Yttrium Barium Copper Oxide) coated conductors have appeared successively [1,2]. Compared with 1G HTS tape, through the improvement of manufacturing process, YBCO HTS tapes have the advantages of higher current density, lower alternating current (AC) loss and lower theoretical cost [3–7], which provide an opportunity for the high power application of superconducting devices applied in the field of electric power.

When current exceeds critical current, superconducting tape will automatically quench and switch to resistance state, which can limit short circuit fault current in power system. Based on the quenching characteristics of YBCO material, resistance type superconducting fault current limiter (R-SFCL) has advantages of simple structure, automatic trigger and fast response. Therefore, R-SFCL can be widely applied to limit short circuit fault current in multiple application scenarios, such as ship power grid [8], railway direct current (DC) traction systems [9], microgrid system [10,11], DC power grid [12–14] and so on.

However, HTS tape is very sensitive to magnetic field, temperature field and stress field. The complex working environment formed by multiple physical fields has a great influence on the homogeneity and stability of superconducting tape. Therefore, the applications of R-SFCL in power system requires the manufacture of superconducting tape with excellent performance and obvious quenching characteristics. It is necessary to focus on studying the quenching characteristic of YBCO material to realize the engineering application of R-SFCL. At present, many scholars have conducted extensive and in-depth research on the different characteristics of YBCO superconducting tape including quenching behavior under AC or DC overcurrent [15–22], quench propagation [23–26], quenching recovery [27–30], thermal and mechanical properties [31,32], minimum quench energy [33,34], maximum operating condition [35], and the effect of different materials on quenching behavior [36–38], which provide a solid foundation for the use of superconducting tape in design and fabrication of R-SFCL. However, there are few studies on common characteristics of quenching resistance of 2G HTS tape under AC and DC overcurrent, which is necessary and basic for R-SFCL design and research.

In addition, it is very difficult to carry out quenching test on a large-scale DC R-SFCL prototype in practical engineering applications before installed in large capacity voltage source converter (VSC) or modular multilevel converter (MMC) high voltage direct current (HVDC) projects. The transmission power of the HVDC projects usually reach hundreds of megawatts. It is very risky to carry out short circuit fault for testing R-SFCL in the existing DC system projects. Meanwhile, the economic and technical cost of manufacturing large-capacity DC test platform for R-SFCL is very high. On the contrary, the technology of large-scale AC experiment platform is mature and its cost is lower. At present, with the support of national key research and development plan, a large-scale DC R-SFCL is being developed in China, which is planned to be installed in Nanao's ±160 kV MMC three-terminal HVDC system. If the common characteristics and internal relations of 2G HTS tape under different type overcurrent is found, the AC overcurrent experimental results of the R-SFCL can be used to evaluate quenching resistance of R-SFCL under DC overcurrent.

In this study, a kind of YBCO tape produced by Shanghai Superconductor Company for R-SFCL was tested under AC and DC overcurrent. By comparing the experimental results of AC and DC overcurrent test, the common characteristics and laws of quenching resistance variation were found under different overcurrent in 77 K liquid nitrogen immersion environment. The same conclusions were obtained through the simulation of YBCO three-dimensional (3D) finite element model. Based on the findings, a new estimation method is proposed, which can accurately reflect and estimate the quenching resistance within a certain range, whose feasibility and availability was verified by comparing with experimental results. Furthermore, it provides theoretical basis and experimental support for AC equivalent test method before a large-scale DC R-SFCL connected to DC power grid.

2. 2G HTS Tape

The YBCO superconducting tapes used in the experiment are produced by Shanghai Superconductor Company (Shanghai, China) [39], which are mainly adopted in R-SFCL. In general, the typical structure of YBCO tape consists of copper stabilizer, silver stabilizer, YBCO layer and Hastelloy substrate. In addition, a buffer layer exists between YBCO layer and Hastelloy substrate.

The 2G HTS tape is 12 mm wide with a two-core configuration, including two typical structures of YBCO tape with copper stabilizer on both side, which are stacked back-to-back and encapsulated with stainless steel layer, as shown in Figure 1a,b (buffer layer is omitted). This structure can increase self-field critical current, which is suitable for power systems with high current. The corresponding basic parameters of the 2G HTS are listed in Table 1. As shown in Figure 1c, the middle 10 cm area of tape samples are selected for voltage measurement in AC and DC overcurrent tests.

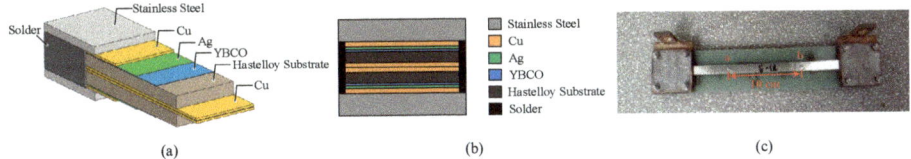

Figure 1. A 2G HTS tape produced by Shanghai Superconductor Company: (**a**) 3D structural diagram; (**b**) sectional view; and (**c**) the sample prepared for experiment.

Table 1. Material parameters of 2G HTS tape used in experiment.

Parameters	Value
Width of tape	12 mm
Thickness of stainless steel layer	80 μm
Thickness of Cu layer	3 μm
Thickness of Ag layer	1.5 μm
Thickness of YBCO layer	1 μm
Thickness of Hastelloy substrate layer	50 μm
Insulation	No
Unit resistance at room temperature	58.42 mΩ/m
Self-field critical current Ic at 77 K	500 A
Effective measurement length of Sample	10 cm

3. AC and DC Overcurrent Experiment

In a strict sense, a current whose magnitude and direction change periodically with time is defined as AC current, while a current whose magnitude and direction remain constant is defined as a DC current. However, in the actual system short circuit fault, the overcurrent will present irregular, even drastic changes. Therefore, in this paper, we define AC and DC overcurrent in a broad sense. The overcurrent in constant direction is DC overcurrent, and the overcurrent in direction varying with time is AC overcurrent. The two kinds of overcurrent analyzed in the experiments are shown in Figure 2.

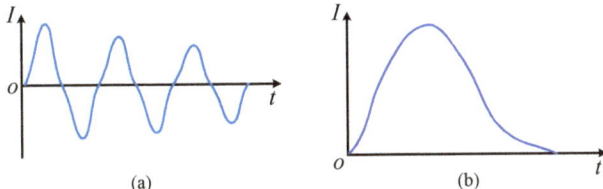

Figure 2. Two type basic waveforms in the experiment: (**a**) AC overcurrent; and (**b**) DC overcurrent.

3.1. AC and DC Overcurrent Experiment Platforms

The AC and DC overcurrent test platforms were, respectively, built for studying quenching characteristics of YBCO tape, as shown in Figures 3 and 4, and their basic corresponding parameters are listed in Table 2. The parameters setting of AC and DC overcurrent experiments are listed in Table 3. The YBCO tape to be tested was immersed in liquid nitrogen and cooled to a superconducting state at 77 K.

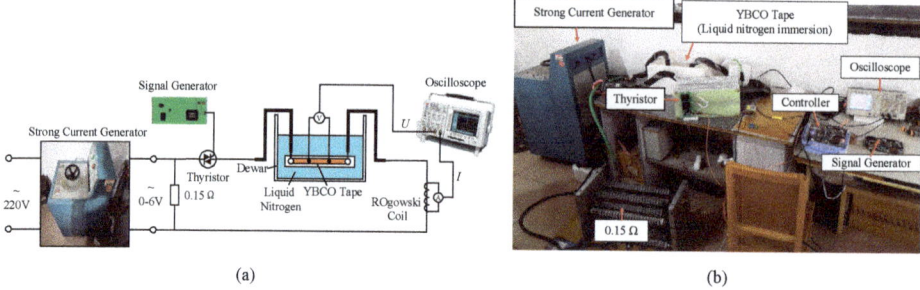

Figure 3. AC overcurrent test platform: (**a**) circuit diagram of test platform; and (**b**) physical map of test platform.

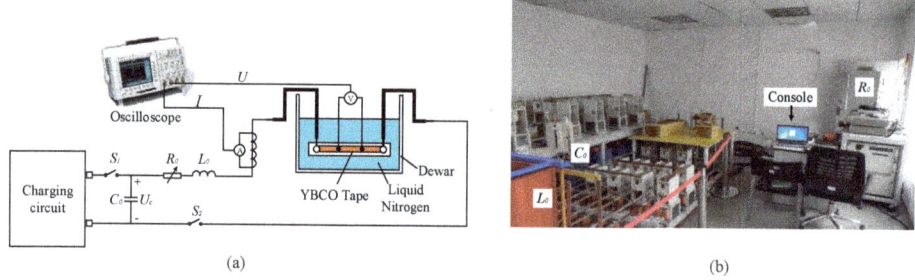

Figure 4. DC overcurrent test platform: (**a**) circuit diagram of test platform; and (**b**) physical map of test platform.

Table 2. Basic parameters of AC overcurrent test platform.

AC Test Platform		DC Test Platform	
Parameters	Value	Parameters	Value
Input voltage	220 V	Output voltage U_C	1000–2200 V
Frequency	50 Hz	Capacitance C_0	8 mF
Output voltage	4–6 V	Inductance L_0	5 mH
Output current	0–3 kA	Resistance R_0	0.2 Ω/0.5 Ω
Duration	50–80 ms	Duration	20–23 ms

Table 3. The parameters setting of AC and DC overcurrent tests.

Number	I_{Pmax} (A)		I_{Pmax}/I_C		Duration (ms)	
AC/DC	AC	DC	AC	DC	AC	DC
No. 1	860	910	1.72	1.82	50	22
No. 2	1060	1260	2.12	2.52	60	23
No. 3	1200	1400	2.4	2.8	60	20
No. 4	1420	1580	2.84	3.16	70	20
No. 5	1580	1620	3.16	3.24	80	22
No. 6	1620	1860	3.24	3.72	60	22
No. 7	1780	1980	3.56	3.96	60	22
No. 8	1920	2020	3.84	4.04	75	22

The AC overcurrent test platform was equipped with a 20 kVA strong current generator, which can turn the 220 V power supply into a 6 V high-power current source with a maximum output of 3 kA. Output capacity was adjusted by strong current generator. The on–off state of thyristor decides duration of AC overcurrent, which was controlled by the signal generator and controller.

The DC overcurrent test platform was the series circuit of resistance R, inductance L and capacitance C (RLC). The RLC series circuit was used to approximately equivalent and simulate the capacitor discharging process of VSC and MMC systems in the DC side short-circuit fault [40,41]. When S_1 is in the closed state and S_2 is in the disconnected state, charging circuit begins to charge the capacitor until the voltage U_C reaches a set value. When S_1 is disconnected and S_2 is closed, R_0, L_0 and C_0 form a closed loop and generate a DC overcurrent for the test of YBCO tape. Peak value, rising slope and duration of DC overcurrent waveform vary with adjusting U_C, L_0 and R_0.

3.2. The results and Analysis of AC Overcurrent Experiment

Eight groups of AC overcurrent tests on YBCO tape were carried out by adjusting the amplitude and duration of overcurrent. The amplitude of overcurrent gradually increased from 860 A to 1920 A. The range of overcurrent duration was 50–80 ms. The eight-group experimental results are shown in Figure 5. The AC experimental analysis is summarized in Table 4.

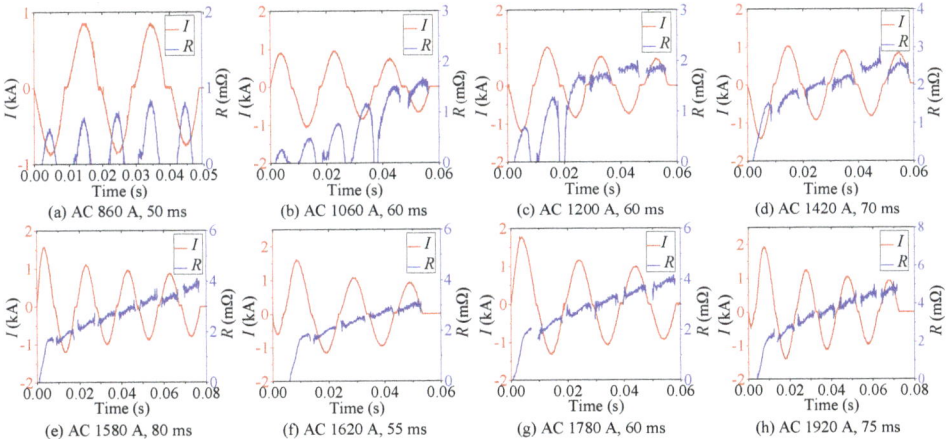

Figure 5. The corresponding experimental results of AC overcurrent test with different amplitude of AC overcurrent.

Table 4. The summary of AC overcurrent experiments.

Resistance Type	Waveform	Results	Characteristics	State
Half-wave type		AC No. 1	Recovery after quenching	Partial resistive
Half-wave and half-incremental curve type		AC No. 2 AC No. 3	Initial stage: Recovery after quenching; Later stage: continuously quenching	Transition: Partial resistive state transitions to normal state
Incremental curve type		AC No. 4 AC No. 5 AC No. 6 AC No. 7 AC No. 8	Continuously quenching to normal state	Normal

As shown in Figure 5a, continuous half-wave type quenching resistance suggests that the quenching state of superconducting tape cannot be maintained, which is unstable and discontinuous. With AC overcurrent decreasing, the superconducting tape returns back to superconducting state and quenching resistance tends to 0 Ω. Therefore, this quenching state is inadequate. At this time, the state of superconducting tape is defined as partial resistive state.

As shown in Figure 5b,c, there are two types resistance waveform under this AC overcurrent level, continuous half-wave type and incremental curve type. In initial stage, superconducting tape recovers rapidly after quenching. Then, superconducting tape begins to quench continuously and transits from fluctuation to continuous increment. It indicates that the quenching degree of superconducting tape gradually deepens and further transits to normal state. This quenching stage of superconducting tape is defined as transition stage, where the state of superconducting tape starts to transit from partial resistive state to normal state.

As shown in Figure 5d–h, incremental curve type quenching resistance reflects that superconducting tape responds to quench continuously without fast recovering under this AC overcurrent level. The quenching state in superconducting tape is adequate and stable, which is considered as complete quenching. It indicates that the superconducting tape is in normal state.

3.3. The Results and Analysis of DC Overcurrent Experiment

The eight-group DC experimental results are presented in Figure 6. The amplitude of overcurrent gradually increases from 910 A to 2020 A. The range of overcurrent duration is 20–23 ms. The DC experimental analysis is summarized in Table 5.

As shown in Figure 6a, half-wave type quenching resistance suggests that the quenching state of superconducting tape cannot always be maintained under this DC overcurrent level. When DC overcurrent decreases, superconducting tape returns to superconducting state rapidly and quenching resistance tends to 0 Ω. Therefore, this quenching state is inadequate, which is considered as incomplete quenching. Similar to the result of AC experiment in Figure 5a, the superconducting tape is in partial resistive state.

As shown in Figure 6b–h, the quenching degree of superconducting tape is strengthened with the peak value I_{Pmax} of DC overcurrent increasing. Because DC overcurrent has no zero crossing before current declines to 0, quenching resistance is incremental curve type. They reflects that superconducting tape quenches continuously without fast recovering under this DC overcurrent level. During this stage, the superconducting tape completely quenches to normal state and presents normal resistance.

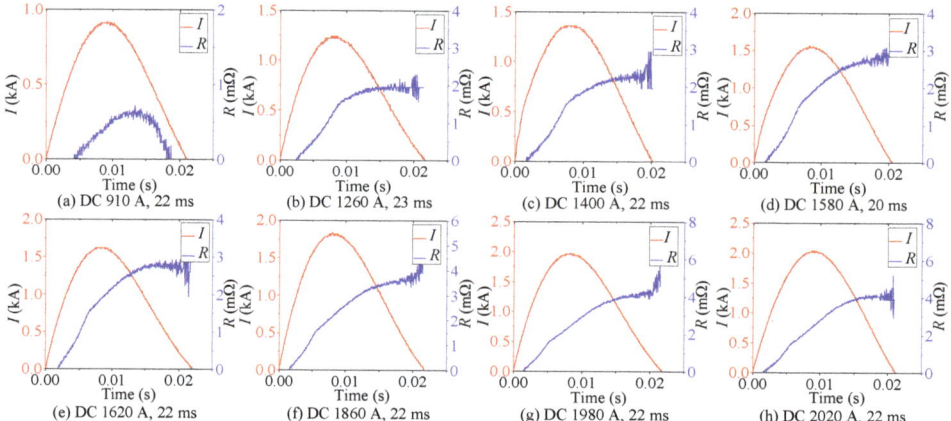

Figure 6. The corresponding experimental results DC overcurrent test with different amplitude of DC overcurrent.

Table 5. The summary of DC overcurrent experiments.

Resistance Type	Waveform	Results	Characteristics	State
Half-wave type		DC No. 1	Recovery after quenching	Partial resistive
Incremental curve type		DC No. 2 DC No. 3 DC No. 4 DC No. 5 DC No. 6 DC No. 7 DC No. 8	Continuously quenching to normal state	Normal

3.4. Comparative Analysis of Quenching Resistance under AC and DC Overcurrent

According to the above results of AC and DC overcurrent experiment shown in Figures 5 and 6, the corresponding *R-t* curves of superconducting tape are obtained, respectively, shown in Figure 7a,b. The *R-t* curves reflect resistance varying with time under different overcurrent. With the increase of overcurrent amplitude, the quenching degree of superconducting tape is enhanced. However, the rising slope and steady-state value of *R-t* curves vary with waveform type, amplitude and duration of impact current. Therefore, *R-t* relationship cannot be used to describe the variation of quenching resistance in superconducting tape.

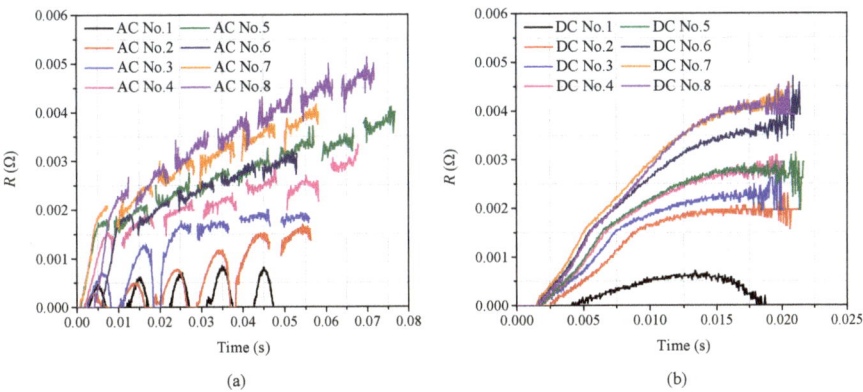

Figure 7. The *R-t* curves of superconducting tape under different overcurrent: (**a**) the *R-t* curves under AC overcurrent; and (**b**) the *R-t* curves under DC overcurrent.

Temperature plays a key role in quenching resistance and quenching degree in superconducting tape. In a certain cryogenic medium, joule heat accumulation is the fundamental cause of temperature rise. Without considering the variation of cryogenic medium, the main factor affecting the temperature of superconducting tape should be the joule heat generated during the quenching process, which is the result of comprehensive effect from quenching resistance, current and quenching duration. The generated joule heat *Q* of the superconducting tape varying with time can be calculated by Equation (1), whose unit is joule (J).

$$Q(t_2) = \int_{t_1}^{t_2} I^2 R(t) \, dt \quad (1)$$

where t_1 is the initial time of quenching, $R(t)$ is quenching resistance corresponding to time, and the range of time *t* is $[t_1, t_2]$. According to sixteen-group data of experimental results, *R-Q* curves of per meter superconducting tape are obtained, where dependent variable *R* is the quenching resistance of per meter superconducting tape and independent variable Q is the generated joule heat of per meter superconducting tape, as shown in Figure 8.

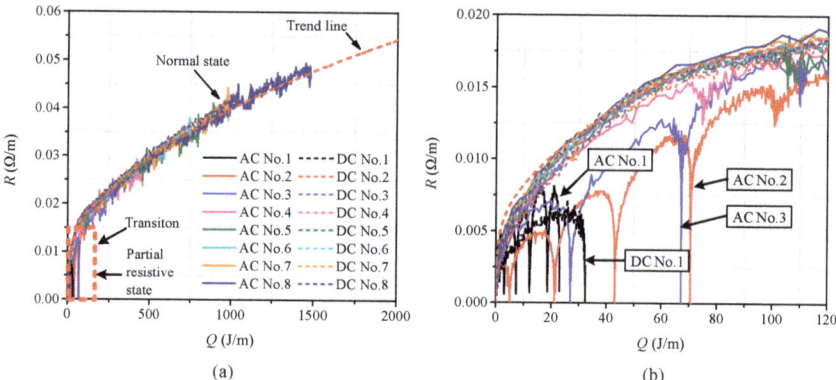

Figure 8. The R-Q curves of superconducting tape under AC and DC overcurrent: (**a**) the comparison diagram of R-Q; and (**b**) partial enlarged drawing of the R-Q curves.

Figure 8a is the comparison diagram of R-Q curves with different overcurrent, and Figure 8b is the partial enlarged drawing of Figure 8a. It is quite clear that twelve R-Q curves obtained are basically overlapping and their trend line are consistent under AC and DC overcurrent except the results of DC No. 1, AC No. 1, AC No. 2 and AC No. 3. According to the classification in Sections 3.2 and 3.3, DC No. 1 and AC No. 1 belong to partial resistive stage of superconducting tape, while AC No. 2 and No. 3 belong to transition stage of superconducting tape.

Therefore, the experimental results present a common trait. When YBCO superconducting tape completely quenches in a cryogenic medium, though superconducting tape is subjected to different overcurrent, the quenching resistance R has an intrinsic fixed relationship with the generated joule heat Q in superconducting tape. In other words, Q is the deciding factor of quenching resistance, while overcurrent is the only inducing factor of quenching resistance. The preliminary finding suggests that, during the complete quenching stage of superconducting tape, quenching resistance can be described by a fixed R-Q curve, which can be obtained from experiments in comprehensive consideration of current, resistance and duration. Therefore, R should satisfy an incremental relationship with Q when superconducting tape tends to complete quenching, described by Equation (2), which also can be reflected by an R-Q curve.

$$R \propto Q(I, R, t) \tag{2}$$

Based on experimental phenomena of superconducting tapes, the quenching resistance characteristics and laws of superconducting tapes are summarized. However, experimental data alone are insufficient to provide adequate physical explanations for the phenomena and laws. Therefore, as presented in Section 4, commercial finite element software COMSOL Multiphysics was adopted to establish 3D simulation model of 2G HTS tape for exploring the criteria of quenching resistance classification, validating the R-Q curve and providing corresponding physical explanations.

4. Simulation Study on 2G HTS Tape Quenching under AC and DC Overcurrent

4.1. 3D Finite Element Model of 2G HTS Tape

In software COMSOL Multiphysics, according to the basic parameters of the superconducting tape listed in Table 1, the 3D multilayer structure of the 2G HTS tape was established as superconducting tape domain, whose surface is covered with liquid nitrogen domain. A combination of custom partial differential equation module (PDE module) and heat transfer module were adopted to perform transient coupling calculation of electric field, magnetic field and thermal field. PDE module was used to carry out electromagnetic calculation. Heat transfer module was used to carry out thermal calculation. The two modules were coupled by joule heat and temperature. The basic structure of the simulation model is shown in Figure 9.

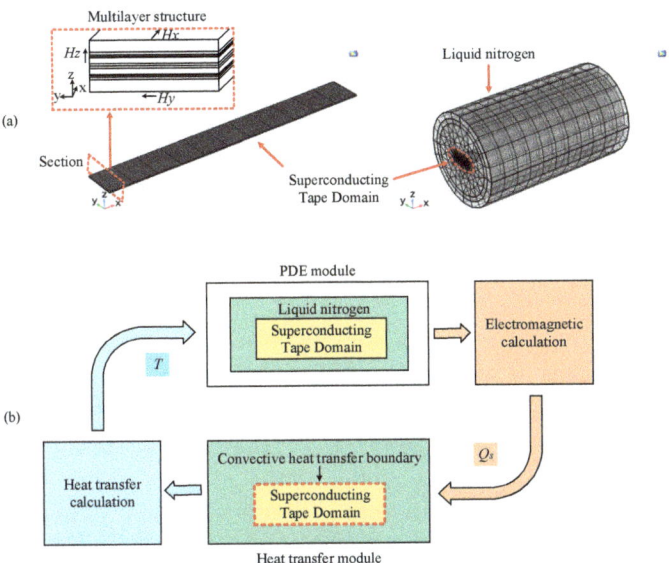

Figure 9. The basic structure of the simulation model: (**a**) geometric structure of superconducting tape; and (**b**) the coupling structure of electromagnetic thermal.

4.2. 3D Electromagnetic Model

The 3D electromagnetic calculation was based on Maxwell equations and H equation [42] with good convergence. The Faraday's law and the ampere-law equation, respectively, are described as Equations (3) and (4).

$$\nabla \times \vec{E} = -\frac{\partial \vec{B}}{\partial t} = -\mu_0 \mu_r \frac{\partial \vec{H}}{\partial t} \tag{3}$$

$$\nabla \times \vec{H} = \vec{J} + \frac{\partial \vec{D}}{\partial t} \tag{4}$$

where $\vec{E}$ is electric field intensity, $\vec{B}$ and $\vec{H}$, respectively, are magnetic induction intensity and magnetic field intensity, and $\vec{J}$ and $\vec{D}$ are conduction current density and displacement current density. Since the conduction current density is much higher than displacement current density in the superconducting tape, the differential term of displacement current with respect to time in Equation (4) can be ignored, thus $\nabla \times \vec{H} = \vec{J}$.

Because $B = \mu_0 \mu_r H$ and $E = rJ$, the partial differential Equation (5) can be obtained according to Equations (3) and (4), where μ_0 is permeability of vacuum, $4\pi \times 10^{-7}$ H/m, μ_r is relative permeability, and r is electric resistivity.

$$\nabla \times \left(r \cdot \nabla \times \vec{H} \right) = -\mu_0 \mu_r \frac{\partial \vec{H}}{\partial t} \tag{5}$$

where $\nabla \times \vec{H} = J_x \vec{e}_x + J_y \vec{e}_y + J_z \vec{e}_z$. J_x, J_y and J_z are the components of J in the x, y, and z directions, which can be expressed by Equations (6).

$$J_x = \frac{\partial H_z}{\partial y} - \frac{\partial H_y}{\partial z}, \quad J_y = \frac{\partial H_x}{\partial z} - \frac{\partial H_z}{\partial x}, \quad J_z = \frac{\partial H_y}{\partial x} - \frac{\partial H_x}{\partial y} \tag{6}$$

Combining with Equations (5) and (6), the governing equation of electromagnetic calculation are obtained as follows.

$$\nabla \cdot \begin{bmatrix} 0 & -E_z & E_y \\ E_z & 0 & -E_x \\ -E_y & E_x & 0 \end{bmatrix} \begin{bmatrix} \vec{e}_x \\ \vec{e}_y \\ \vec{e}_z \end{bmatrix} + \frac{\partial}{\partial t} \begin{bmatrix} \vec{e}_x & \vec{e}_y & \vec{e}_z \end{bmatrix} \begin{bmatrix} \mu_0\mu_r & 0 & 0 \\ 0 & \mu_0\mu_r & 0 \\ 0 & 0 & \mu_0\mu_r \end{bmatrix} \begin{bmatrix} H_x \\ H_y \\ H_z \end{bmatrix} = 0 \tag{7}$$

where $\nabla = [\frac{\partial}{\partial x}, \frac{\partial}{\partial y}, \frac{\partial}{\partial z}]$. $\vec{e}_x, \vec{e}_y,$ and $\vec{e}_z$ are, respectively, the unit direction vector of x, y, and z directions. E_x, E_y, and E_z are the components of E in the x, y, and z directions. H_x, H_y, and H_z are the components of H in the x, y, and z directions, which are set as dependent variables in PDE module.

Overcurrent was used as model excitation in simulation. As a composite conductor, each layer of the superconducting tape was assumed to be parallel structure. Therefore, the current constraint condition of the model should meet Equation (8), which is load on the cross section of superconducting tape by pointwise constraint.

$$I_{input} = I_{SC} + I_{Cu} + I_{Ag} + I_{Sub} + I_{St} \tag{8}$$

where I_{input} is total current excitation, I_{SC} is the current in YBCO layer, I_{Cu} is the current in copper layer, I_{Ag} is the current in silver layer, I_{Sub} is the current in Hastelloy substrate, and I_{St} is the current in stainless steel layer.

In the quenching model of superconducting tape, the calculation of YBCO resistivity is crucial; a piecewise formula for YBCO resistivity was adopted, which is expressed as Equations (9). Critical temperature T_c is used as criterion for state transition. When $T \leq T_c$, the superconducting tape is considered as incomplete quenching, whose YBCO resistivity adopts the classical method of parallel equivalent electrical circuit [43]. When $T > T_c$, the superconducting tape is considered as complete quenching, whose YBCO resistivity adopts constant resistivity r_{norm}.

$$r_{YBCO} = \begin{cases} \frac{(r_1+r_2+r_0)r_{norm}}{r_0+r_1+r_2+r_{norm}} & T \leq T_c \\ r_{norm} & T > T_c \end{cases} \tag{9}$$

where r_0 is residual resistivity [44], r_1 is expressed as Equation (10), r_2 is expressed as Equation (11). According to the authors of [42], n_1, n_2 and E_0 are fitting parameters of two power-law relations, and k can be calculated based on $r_1(3J_c) = r_2(3J_c)$.

$$r_1 = \begin{cases} 0 & |J| < J_c \\ \frac{E_0}{|J|}\left(\frac{|J|}{J_c} - k\right)^{n_1} & |J| \geq J_c \end{cases} \tag{10}$$

$$r_2 = \begin{cases} 0 & |J| < kJ_c \\ \frac{E_0}{|J|}\left(\frac{|J|}{J_c} - k\right)^{n_2} & |J| \geq kJ_c \end{cases} \tag{11}$$

According to the authors of [45], when the temperature T of YBCO layer is below critical temperature T_c, the critical current density J_c of YBCO material is affected by temperature T. When T is above T_c, J_c is considered as 0 A/m^2. The relationship can be expressed by the improved typical Equation (12) in [43].

$$J_c(T) = \begin{cases} J_{c0}\left(\frac{T_c-T}{T_c-T_{ref}}\right)^{\alpha} & (T_{ref} < T < T_c) \\ 0 & (T_c \leq T) \end{cases} \tag{12}$$

where T_{ref} is initial ambient temperature or cryogenic medium temperature of the superconducting tape. J_{c0} is critical current density corresponding to T_{ref}. In this simulation, superconducting tapes were cooled by liquid nitrogen immersion. Therefore, T_{ref} was set to 77 K. The relevant simulation parameters of YBCO resistivity are listed in the Table 6.

Table 6. Simulation parameters of YBCO resistivity.

Parameters	Value	Parameters	Value
r_0	1×10^{-14} Ω·cm	n_1	2.8
r_{norm}	2.5 μΩ·cm	n_2	22
E_0	0.5 V/cm	k	1.92
T_c	90 K	α	1.5
J_{c0}	1.9×10^{10} A/m²	T_{ref}	77 K

4.2.1. Heat transfer Model

In heat transfer calculation, because superconducting tape was immersed in liquid nitrogen, which is without obvious flowing, a heat balance Equations (13) in the module of Heat Transfer in Solids in COMSOL Multiphysics was used to calculate the thermal characteristics of the entire superconducting tape region including YBCO layer, Cu layer, Ag layer, stainless steel and Hastelloy substrate.

$$\begin{cases} q_s = \rho c \frac{\partial T}{\partial t} + \nabla \cdot q \\ q_s = E_x J_x + E_y J_y + E_z J_z \\ q = -k \nabla T \end{cases} \quad (13)$$

where q_s is volume power density, with unit W/m³; ρ is mass density, with unit kg/m³; c is specific heat capacity, with unit J/(kg·K); q is conduction heat flux, with unit W/m²; and k is heat transfer coefficient, with unit W/(m²·K).

In the simulation, the initial temperature of thermal solution domain was set to 77 K, and the boundary condition of the superconducting tape surface was set to Heat Flux. To simulate the heat exchange process between superconducting tape surface and liquid nitrogen, a heat transfer coefficient curve in Heat Flux was set on the interface between equivalent superconducting tape and liquid nitrogen [43,46,47], which is a common method to simulate heat exchange process between liquid nitrogen and superconducting tape surface at different temperatures. As shown in Figure 10, this is a heat transfer coefficient curve of liquid nitrogen including free convection, nuclear boiling, transition boiling and film boiling [48].

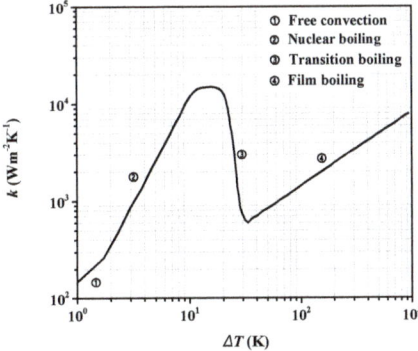

Figure 10. Heat transfer coefficient curve of liquid nitrogen.

4.3. Verification of the 3D Finite Element Model

As shown in Figure 11a–d, four-group quenching data of superconducting tape under different AC overcurrent were selected for comparison between experiment and simulation, covering three types quenching resistance waveforms of superconducting tapes under AC overcurrent. Figure 11e–h compares the results of experiment and simulation, covering two types quenching resistance waveforms of superconducting tape under DC overcurrent.

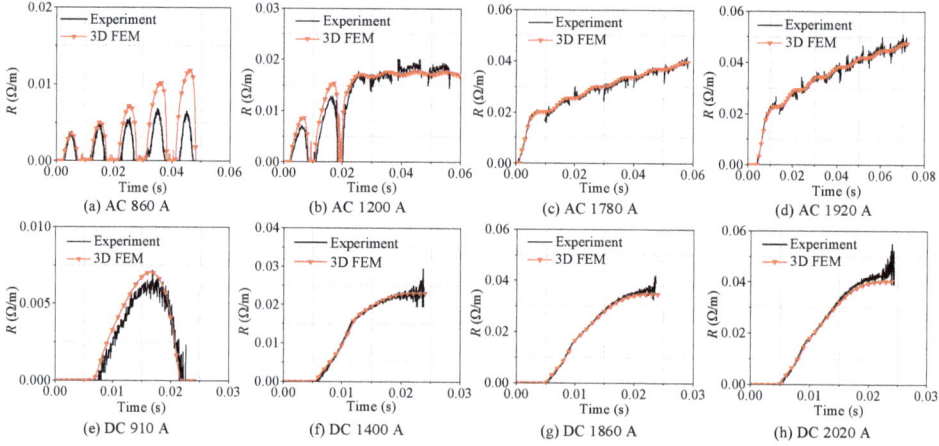

Figure 11. Comparison of experimental and simulation results.

Through waveform comparison between experiment and simulation, it can be found that the simulation phenomena are basically consistent with the classification of experimental results, where the model can correctly describe the different types of quenching resistance under different overcurrent. Therefore, the 3D finite element model can correctly reflect the quenching state of the superconducting tape, and contributes to research the internal evolution process of superconducting tape under AC and DC overcurrent.

4.4. Thermal Characteristics and Current Distribution Characteristics of 2G HTS Tape Quenching

4.4.1. Resistance, Thermal and Current Distribution Characteristics under AC Overcurrent

To analyze the transient characteristics of superconducting tape under AC overcurrent and obtain corresponding common laws, it was necessary to select and analyze the three typical quenching resistance waveforms of superconducting tape found in Section 3.2 under AC overcurrent, including half-wave type, half-wave and half-incremental curve type and incremental curve type. Therefore, the simulation results with AC 860 A, AC 1200 A and AC 1780 A were selected.

Figure 12 is the simulation results of the superconducting tape quenching under AC 860 A, which represent the characteristics of continuous half-wave type quenching resistance in partial resistive state. According to Figure 12a, when $0 \le t < t_1$, the superconducting tape is in superconducting state with 0 Ω, and current only flows through YBCO layer. When $t_1 \le t$, current begins to exceed the critical current I_C which results in YBCO layer quenching. However, the ratio of real-time resistivity to normal resistivity of YBCO is very low, no more than 0.07. It indicates that the superconducting tape is in partial resistive state. As shown in Figure 12b, during t_1 to 60 ms, the temperature of YBCO layer rises continuously in fluctuation, but the maximum value of temperature is only 88.7 K, which deso not exceed the critical temperature T_C of YBCO layer. Therefore, during the whole quenching process, the quenching resistivity of YBCO layer is in the first stage in Equation (9), which causes the current to be diverted to other layers. The resistivity of YBCO layer fluctuates with the rise and fall of AC overcurrent, which results in the superconducting tape producing the corresponding continuous half-wave type quenching resistance. During initial quenching, the current distribution among material layers is: $I_{YBCO} > I_{Cu} > I_{Ag} > I_{St} > I_{Sub}$. With the temperature of YBCO layer rising, the quenching resistivity of YBCO layer increases, and the ordering of current in proportion is: $I_{Cu} > I_{YBCO} > I_{Ag} > I_{St} > I_{Sub}$.

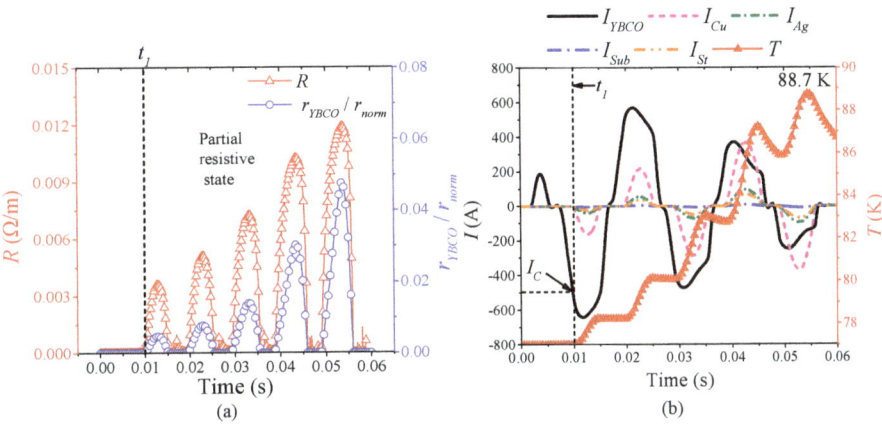

Figure 12. The simulation analysis with AC 860 A overcurrent: (**a**) the quenching resistance characteristics; and (**b**) the thermal characteristics and current distribution characteristics.

Figure 13 is the simulation results of the superconducting tape quenching under AC 1200 A, which represents the characteristics of continuous half-wave and half-incremental curve type quenching resistance in transition state. According to Figure 13a, when $0 \leq t < t_1$, the superconducting tape is in superconducting state with 0 Ω, current only flow through YBCO layer. When $t_1 \leq t < t_2$, the ratio of real-time resistivity to normal resistivity of YBCO is very low, not reaching 1. Therefore, the quenching characteristics and current distribution characteristics of this condition are the same as those in partial resistive state. According to Figure 13b, when $t_2 \leq t$, the temperature of YBCO layer exceeds the critical temperature T_C. Its maximum temperature is 92.3 K, which results in the quenching resistivity of YBCO transformed into constant resistivity in Equation (9). At this time, the ratio of real-time resistivity to normal resistivity of YBCO is 1, which indicates that YBCO layer transitions from partial resistive state to normal state. Therefore, current mainly flows through other layers except YBCO layer. During this period, the characteristics of current distribution is: $I_{Cu} > I_{Ag} > I_{St} > I_{Sub} > I_{YBCO}$. The superconducting tape further presents a composite conventional metal resistivity. However, with the decline of AC overcurrent, the superconducting tape gradually reaches thermal equilibrium. The temperature fluctuation near 90 K causes the quenching degree of YBCO layer to be in a critical state, fluctuating between partial resistive state and normal state.

Figure 14 is the simulation results of the superconducting tape quenching under AC 1780 A, which represent the characteristics of continuous incremental curve type quenching resistance in normal state. In this condition, the duration of partial resistance state is very short. Current rapidly exceeds the critical current of YBCO layer. Temperature correspondingly exceeds critical temperature 90 K within first current half wave, thereby resulting in YBCO layer transformed into normal state within about 5 ms, as shown in Figure 14. With temperature continuously rising, the superconducting tape presents corresponding incremental curve type quenching resistance. The maximum temperature of YBCO layer is 174 K. During most of quenching process, the current distribution in superconducting tape is mainly: $I_{Cu} > I_{Ag} > I_{St} > I_{Sub} > I_{YBCO}$.

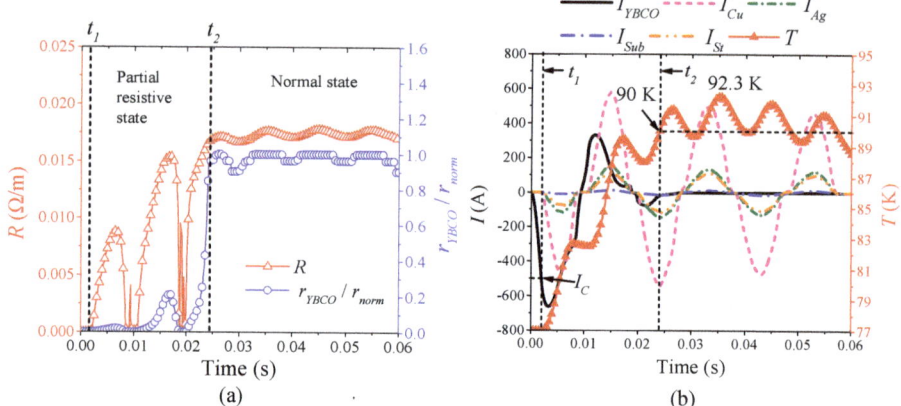

Figure 13. The simulation analysis with AC 1200 A overcurrent: (**a**) the quenching resistance characteristics; and (**b**) the thermal characteristics and current distribution characteristics.

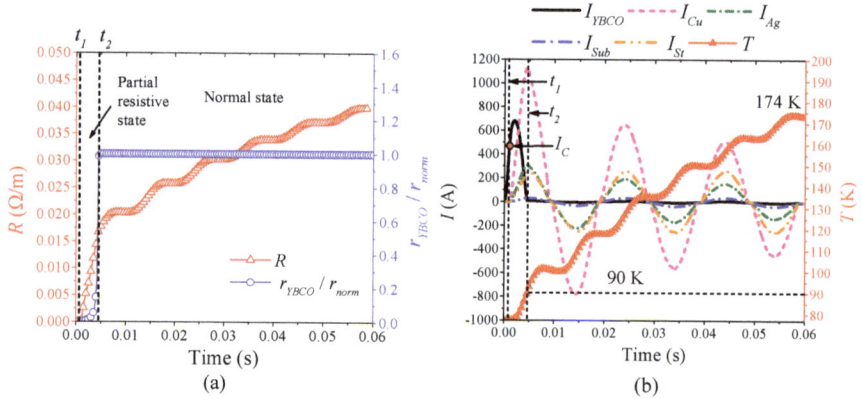

Figure 14. The simulation analysis with AC 1780 A overcurrent: (**a**) the quenching resistance characteristics; and (**b**) the thermal characteristics and current distribution characteristics.

4.4.2. Resistance, Thermal and Current Distribution Characteristics under DC Overcurrent.

The two typical quenching resistance waveforms of superconducting are found under DC overcurrent in Section 3.3 including half-wave type and incremental curve type. Therefore, the simulation results with DC 910 A and DC 2020 A were selected to analyze the transient characteristics of superconducting tape under DC overcurrent to obtain the corresponding common laws.

Figure 15 is the simulation results of the superconducting tape quenching under DC 910 A, which represent the characteristics of half-wave type quenching resistance in DC partial resistive state, whose quenching characteristics are same as those of AC partial resistive state. When $0 \leq t < t_1$, the superconducting tape is in superconducting state with 0 Ω. Therefore, DC overcurrent only flows through YBCO layer. When $t_1 \leq t$, DC overcurrent exceeds the critical current of YBCO layer, but the temperature is still below the critical temperature 90 K. During the whole quenching process, the maximum temperature of YBCO layer is 82.9 K. Therefore, the quenching resistivity of YBCO layers is far below normal resistivity, as shown in Figure 16a, resulting in the superconducting tape recovering quickly with DC overcurrent attenuation, thereby forming half-wave type quenching resistance. The current distribution characteristic of superconducting tape is: $I_{YBCO} > I_{Cu} > I_{Ag} > I_{St} > I_{Sub}$ in Figure 15b).

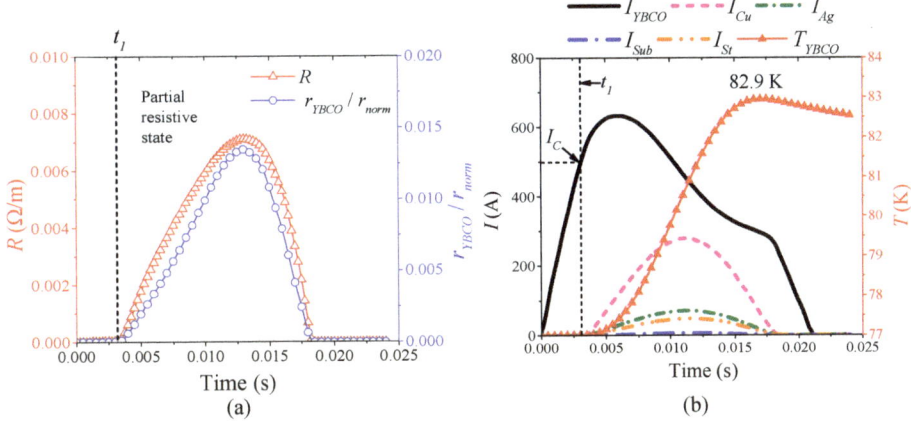

Figure 15. The simulation analysis with DC 910 A overcurrent: (**a**) the quenching resistance characteristics; and (**b**) the thermal characteristics and current distribution characteristics.

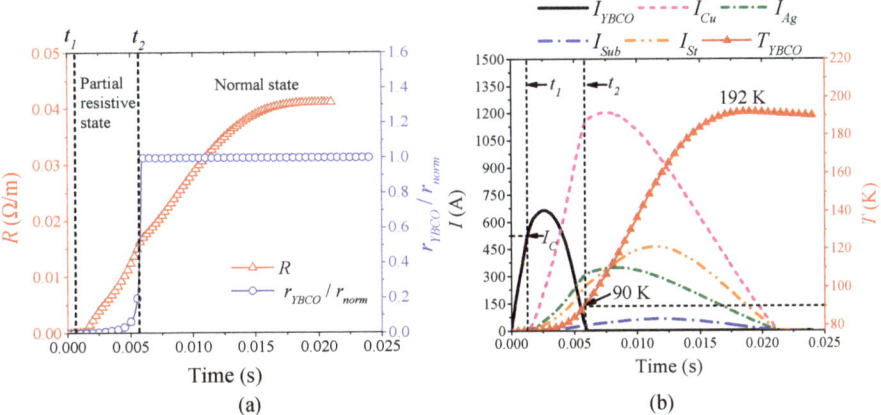

Figure 16. The simulation analysis with DC 2020 A overcurrent: (**a**) the quenching resistance characteristics; and (**b**) the thermal characteristics and current distribution characteristics.

Figure 16 is the simulation results of the superconducting tape quenching under DC 2020 A, which represent the characteristics of incremental curve type quenching resistance in DC normal state. The quenching characteristics are the same as those of AC normal state. Because the temperature of YBCO layer exceeds the critical temperature and increases continuously, the resistivity of YBCO layer quickly transits from partial resistive state to normal state, as shown in Figure 16a. Therefore, current only flows through other layers except YBCO layer. The equivalent quenching resistance of the superconducting tape is composite resistance of metal materials depending on the temperature, which presents incremental curve type with temperature rising. Although DC overcurrent starts to decay after the first crest, the temperature is maintained at 192 K. Therefore, the superconducting tape remains at normal resistivity state, whose recovery is very slow. In this condition, current distribution characteristics is: $I_{Cu} > I_{Ag} > I_{St} > I_{Sub} > I_{YBCO}$ (Figure 16b).

4.5. Consistency Analysis of R-Q Curve

According to the simulation results shown in Figure 11, Equation (2) is adopted to plot the R-Q curves shown in Figure 17, which present the same phenomenon as the experimental results. When superconducting tape is in normal state in a cryogenic medium, the variation tendency of R-Q curves under different AC and DC overcurrent is consistent, while the partial resistive state and transition phenomena do not appear.

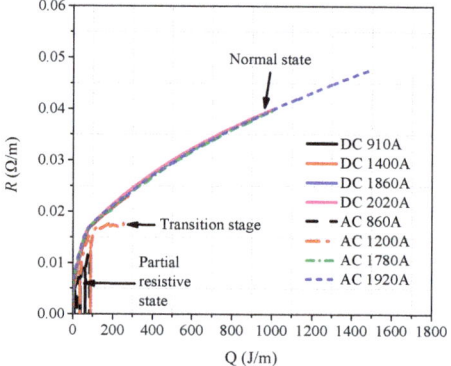

Figure 17. The simulation results of R-Q curves.

To explain the phenomena and analyze the consistency in R-Q curves in normal state, the heat transfer process of each simulation result is plotted in Figure 18, including heat power P_{heat} generated by superconducting tape, cooling power P_{cool} of liquid nitrogen on superconducting tape surface and temperature T of superconducting tape. In the consideration of the heat exchange process between superconducting tape and liquid nitrogen, accumulated joule heat Q_1 can be calculated by Equation (14).

$$Q_1(t_2) = \int_{t_1}^{t_2} (P_{heat} - P_{cool}) dt \quad (14)$$

$$Q_1 = cm\Delta T \quad (15)$$

where c is specific heat capacity, m is mass, and ΔT is equal to $T-T_{ref}$. According to Equation (15), accumulated joule heat Q_1 directly causes the temperature T to rise, which can further aggravate quenching.

When superconducting tape is in partial resistive state or transition, the temperature of superconducting tape is below 100 K. At this time, liquid nitrogen is in free convection, nuclear boiling or transition boiling. During this time, the cooling power of liquid nitrogen is relatively high, which greatly weakens the quench of superconducting tape. Therefore, the effect of P_{cool} cannot be ignored, as shown in Figure 18a,b,e. In addition, compared with the heating power of superconducting tape, the cooling power of liquid nitrogen shows obvious hysteresis phenomenon. Therefore, heating power increases and decreases rapidly with the change of overcurrent, resulting in temperature rising first and then dropping. During partial resistive state and transition, R-Q curves without considering P_{cool} cannot correctly reflect the changing rule of quenching resistance, which further validates and explains that the R-Q curves are not consistent at these quenching stages in the experimental results in Figure 8b.

When superconducting tape is in normal state, the surface temperature of the superconducting tape is very high, causing the surrounding liquid nitrogen to be in film boiling. Because the cooling power of liquid nitrogen is much less than the heating power of superconducting tape ($P_{heat} \gg P_{cool}$), the thermal equilibrium is quickly broken by a huge thermal shock. Hence, superconducting tape is approximately in an adiabatic environment, whose heat exchange with liquid nitrogen can be ignored, as shown in Figure 18c,d,f–h. At this time, the superconducting tape is in thermal runaway; quenching resistance increases rapidly with temperature rising. P_{cool} can be omitted, thus the amount of joule heat Q generated by superconducting tape can be approximately equivalent to accumulated joule heat Q_1.

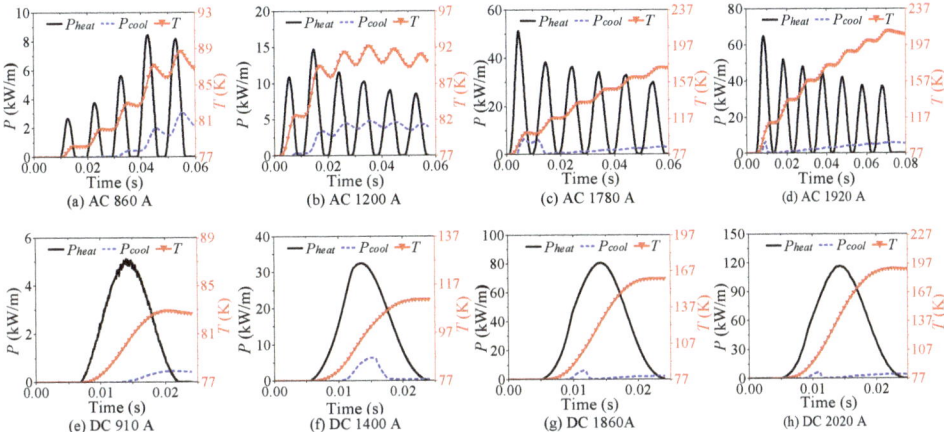

Figure 18. The simulation analysis of heat transfer process of 1 m superconducting tape under AC and DC overcurrent.

In the complete quenching process of superconducting tape, due to $T > T_c$, the quenching resistivity of YBCO layer is normal resistivity r_{norm}. The resistivity of other materials have fixed curves varying with temperature T, as shown in the Appendix A. The equivalent resistivity of the superconducting tape can be calculated by Equation (16).

$$r_{eq} = 1 / \left(\sum_{i=1}^{n} \frac{f_i}{r_i} \right) \tag{16}$$

$$R_{eq} = \frac{r_{eq} \cdot l}{S} \tag{17}$$

where r_i is the resistivity of each material layer. f_i is the volume percentage of each material in the superconducting tape. l is the length of superconducting tape and S is the cross-section area. According to Equations (16) and (17), the equivalent resistance of 1 m superconducting tape varying with temperature is a fixed curve, as shown in Figure 19a.

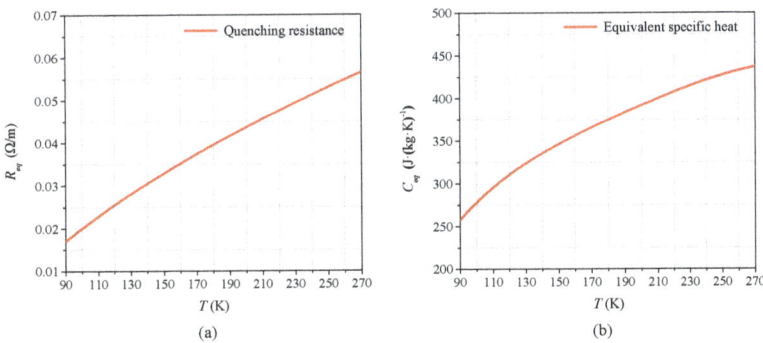

Figure 19. Equivalent material properties: (**a**) the equivalent quenching resistance curve; and (**b**) the equivalent specific heat capacity.

The specific heat capacity c of each material has a fixed curve varying with temperature T, as shown in the Appendix A. The equivalent specific heat capacity c_{eq} of the superconducting tape can be calculated by Equations (18), as shown in Figure 19b.

$$\begin{cases} c_v = 1 / \left(\sum_{i=1}^{n} f_i / (\gamma_i \cdot c_i) \right) \\ c_p = \sum_{i=1}^{n} f_i \cdot \gamma_i \cdot c_i \\ c_{eq} = (c_v + c_p)/2 \end{cases} \quad (18)$$

where c_v is the equivalent specific heat capacity in the vertical direction of the superconducting tape surface, c_p is the equivalent specific heat capacity in the parallel direction of the superconducting tape surface, c_i is the specific heat capacity of each material, and γ_i is the mass density ratio of each material layer to the whole superconducting tape.

With the superconducting tape quenching to normal state, the temperature of superconducting tape is mainly affected by the amount of joule heat generated with time. Therefore, during this condition, the R-Q curves of superconducting tape under different AC and DC overcurrent are consistent, which can be used to reflect the change law of quenching resistance. According to Equations (14)–(18), the theoretical R-Q curve in normal state can be calculated by numerical calculation, which is compared with experiment results and FEM results in Figure 20.

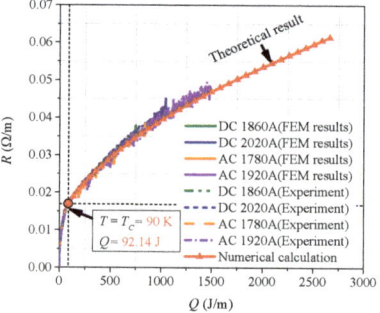

Figure 20. The comparison diagram of R-Q curves from different results.

The comparison in Figure 20 obviously indicates that, when $T > T_C$, the R-Q curves of superconducting tape are basically consistent with the theoretical result. In normal state, critical temperature T_C is the turning point of quenching resistance variation. The accumulated joule heat required is about 92.14 J. Furthermore, it also confirms the correctness of the theoretical analysis of R-Q curves consistency mentioned above. When the superconducting tape enters normal state, it is approximately in an adiabatic environment, and quenching resistance will vary with the accumulated joule heat along a fixed R-Q curve, which is consistent under AC and DC overcurrent.

4.6. The Summary of Quenching Characteristics under AC and DC Overcurrent

Based on the above simulation and experimental analysis, the superconducting tape is divided into three stages: superconducting state, partial resistive state and normal state. Their classified method is shown in Figure 21 and corresponding characteristics are listed in Table 7. Critical current $I_C(T)$ is the criterion of whether or not to quench. Critical temperature T_C is the criterion of quenching degree. The consistency of R-Q curves is satisfied in normal state.

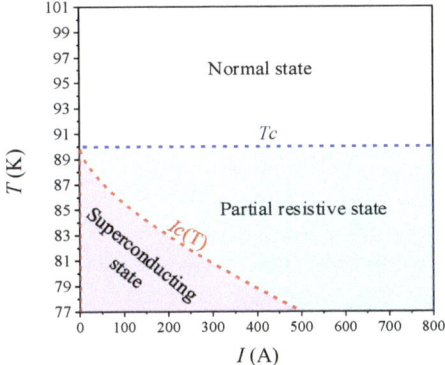

Figure 21. Different stage distribution of superconducting tape immersed in liquid nitrogen.

Table 7. The common quenching characteristics of the superconducting tape.

State		Boundary	Current Distribution Characteristics	YBCO Resistivity	R-Q Consistency
Superconducting state		$I < I_C(T)$	Only flow through superconducting layer	0	/
Quenching	Partial resistive state:	$I > I_C(T)$ $T < T_C$	Initial stage: $I_{YBCO} > I_{Cu} > I_{Ag} > I_{St} > I_{Sub}$ Late stage: $I_{Cu} > I_{YBCO} > I_{Ag} > I_{St} > I_{Sub}$	$\frac{(r_1+r_2+r_0)r_{norm}}{r_0+r_1+r_2+r_{norm}}$	No
	Normal state	$T_C < T$	$I_{Cu} > I_{Ag} > I_{St} > I_{Sub} > I_{YBCO}$	r_{norm}	Yes

5. The Equivalent Estimation of Quenching Resistance under AC and DC Overcurrent

5.1. The Basic Principle of the New Method

Combined with experiment and FEM simulation, a general rule is further verified. When superconducting tape quenches to normal state, immersed in liquid nitrogen environment, the superconducting tape has a correspondence relationship between quenching resistance and the accumulated joule heat. The R-Q curve is consistent under different AC and DC overcurrent. Therefore, a new calculation method of quenching resistance is proposed to estimate and predict variation of quenching resistance when superconducting tape is in normal state. The joule heat adopted in the new method can effectively evaluate the quenching resistance of superconducting tape and ignore the difference of current waveforms.

The method is defined as "R-Q curve method", whose calculation process is shown in Figure 22. In the calculation process, test current is used as input current. I_C is quenching criterion. R_0 is set as initial quenching resistance for initial energy accumulation. Δt is the time step of calculation. When superconducting tape starts to quench, the accumulated joule heat Q in each time step is calculated by discrete calculation, whose corresponding quenching resistance R can be obtained by interpolation calculation based on R-Q curve. In addition, under overcurrent shock, the effect of liquid nitrogen on superconducting tape can be approximately ignored. Therefore, the temperature variation of superconducting tape can be calculated by Equation (19).

$$T = \frac{Q}{cm} + T_{ref} \qquad (19)$$

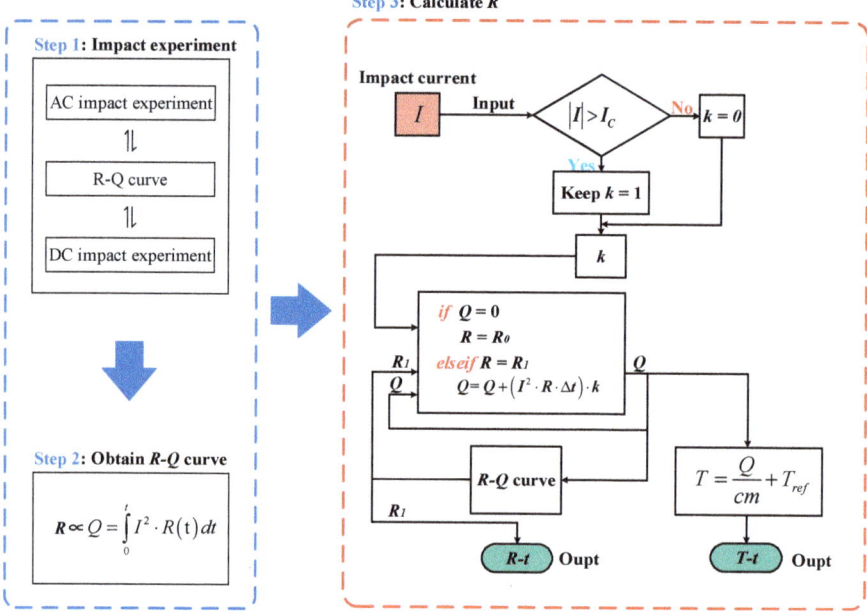

Figure 22. The calculation process of R-Q curve method.

5.2. The Validation of R-Q Curve Method

According to Sections 3 and 4, under AC and DC overcurrent, the *R-Q* curves of superconducting tape obtained have the same trend. Therefore, the experimental data with the most extensive coverage are adopted to be converted into the *R-Q* curve of per meter superconducting tape, as shown in Figure 23, whose data are from AC 1920 A overcurrent test. The simulation parameters setting are listed in Table 8.

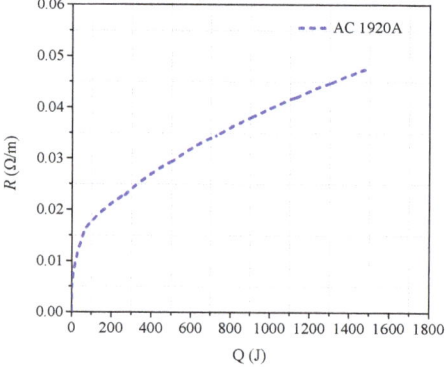

Figure 23. *R-Q* curve required in simulation calculation.

Table 8. Simulation parameters setting of *R-Q* curve method.

Parameters	Value
Critical current, I_C	500 A
Initial quenching resistance, R_0	0.001 Ω
Time step, Δt	5×10^{-5} s
Length of superconducting tape	1 m
Overcurrent	AC 1780 A, AC 1920 A, DC 1860 A, DC 2020 A
Duration	60 ms, 80 ms, 25 ms, 25 ms

According to Figure 24, with comparison between simulations and experiments, the effectiveness of *R-Q* curve method is well verified, which can be used to simulate the variation process of quenching resistance of small-scale superconducting tape in normal state under AC and DC overcurrent. In addition, the *R-Q* curve in the simulation comes from the AC experimental data, but the model of *R-Q* curve method still has good simulation accuracy under DC overcurrent. It also indicates that the quenching resistance of AC and DC overcurrent test can be equivalent by *R-Q* curve in normal state. Meanwhile *R-Q* curve method can calculate temperature change in the quenching process, whose results are basically consistent with the results of FEM but computing speed is faster, as shown in Figure 25.

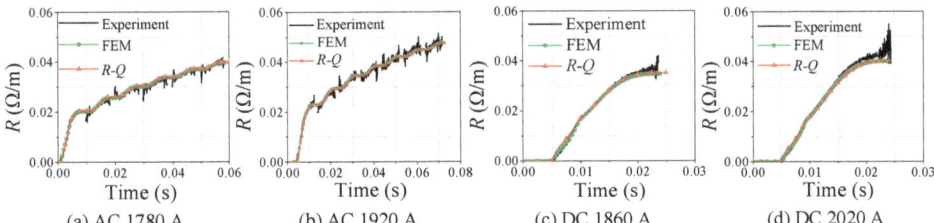

Figure 24. The quenching resistance comparison of FEM method, *R-Q* curve method and experiment.

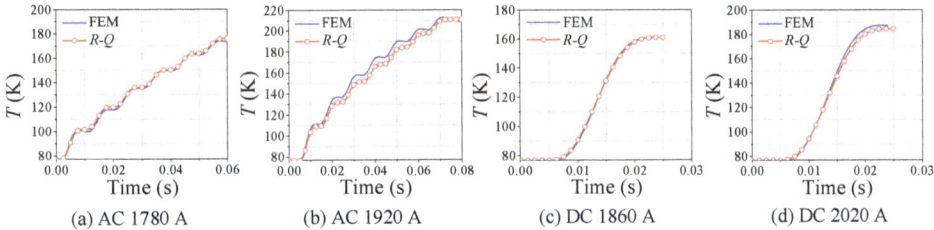

Figure 25. The temperature comparison of FEM and *R-Q* curve method.

Obviously, in terms of estimating quenching resistance of superconducting tape, *R-Q* curve method has a good calculation accuracy in normal state. However, compared with FEM, because of *R-Q* curve method using interpolation calculation while FEM adopting the coupling calculation of multiple physical fields, the model structure of *R-Q* curve method is simpler and faster, as shown in Table 9. However, it is undeniable that the application range of *R-Q* curve method is limited.

Having good simulation precision and calculation speed, the *R-Q* method can be applied to evaluate and predict the quenching resistance of large-scale non-inductive superconducting coil in R-SFCL based on the experimental data of small-scale superconducting tape. The prediction results of the model is compared with the experimental results of the coil to verify this estimation method.

Table 9. The computation time of R-Q curve method and FEM.

Overcurrent	Software	Method	Computation Time
AC 1780A, 60 ms	PSCAD	R-Q	<3 s
	Comsol	FEM	3.78 h
AC 1920A, 75 ms	PSCAD	R-Q	<3 s
	Comsol	FEM	3.85 h
DC 1860A, 22 ms	PSCAD	R-Q	<3 s
	Comsol	FEM	1.57 h
DC 2020A, 22 ms	PSCAD	R-Q	<3 s
	Comsol	FEM	1.58 h

To calculate the quenching resistance of superconducting tape with any length, the model of R-Q curve method is improved in PSCAD, as shown in Figure 26, whose R-Q curve data comes from the experimental results of 10 cm superconducting tape converted to the R-Q curve of 1 m superconducting tape applied in the model. The coefficient K is the length of superconducting tape, which can be set according to the length of experimental sample including the length of superconducting tape in non-inductive superconducting coil.

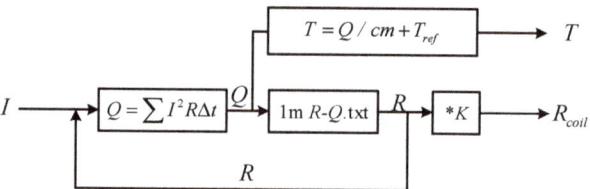

Figure 26. The model of R-Q curve method improved in PSCAD.

The non-inductive superconducting coil is shown in Figure 27, which adopts single 12 mm width superconducting tape winding. Its diameter is 1200 mm and its length of superconducting tape is 136 m. Because the superconducting coil adopts non-inductive design, the effect of magnetic field on superconducting tape can be neglected. According to Figure 28, the peak value of DC overcurrent is 3.7 kA, which results in superconducting coil rapidly entering in complete quenching stage. The inductance of superconducting non-inductive coil can be ignored, but its resistance needs to be considered, where resistance and accumulated joule heat are proportional to the length of superconducting tape. Without considering the influence of coil structure, the quenching resistance of 136 m superconducting tape calculated by R-Q curve method is basically consistent with the experimental results of actual superconducting coil. It indicates that the non-inductive coil structure can effectively promote the uniform quench of the superconducting coil.

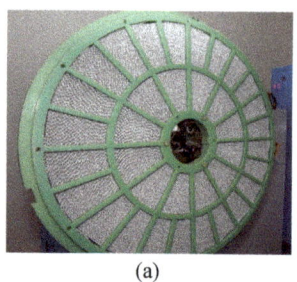

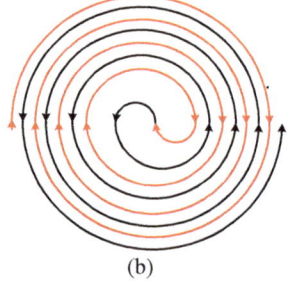

(a)

(b)

Figure 27. The schematic diagram of non-inductive superconducting coil: (a) physical drawing; and (b) structure drawing.

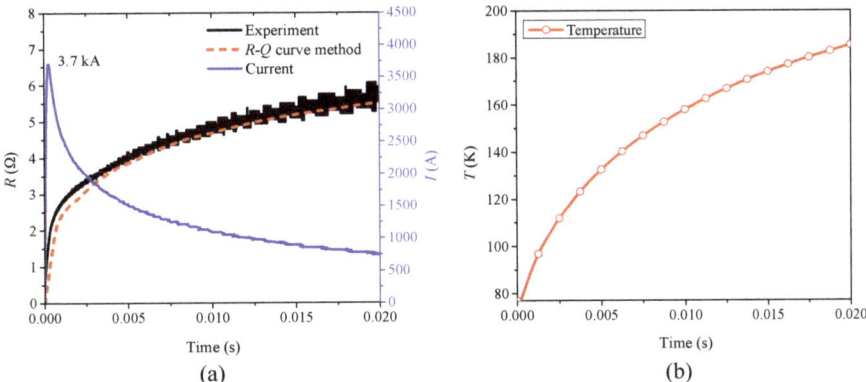

Figure 28. Comparison between the result of *R-Q* curve method and experimental result of 136 m non-inductive superconducting coil: (**a**) quenching resistance; and (**b**) temperature estimation.

In addition, Figure 28 clearly shows that *R-Q* curve method has good simulation accuracy and fast solving speed compared with the experimental results in the time range of tens of milliseconds. The *R-Q* method can rapidly calculate the quenching resistance and average temperature of large non-inductive superconducting coils during complete quenching, only taking a few seconds. Therefore, the method can be improved to predict the dynamic quenching resistance and temperature of superconducting coil for R-SFCL design.

6. Conclusions

In this paper, through the combination of AC and DC overcurrent experiment and simulation study, the quenching characteristics and common laws of YBCO superconducting tape applied in R-SFCL are summarized as follows:

(1) Under AC overcurrent, the quenching resistance of the HTS tape can be divided into three types: half-wave type, half-wave and half-incremental curve type and incremental curve type. Under the DC overcurrent, there are two quenching resistance types of the HTS tape: half-wave type and incremental curve type.

(2) The quench resistance type is closely related to the current and the temperature of the HTS tape. When the current exceeds the critical current and the temperature is lower than the critical temperature of the HTS tape, the HTS tape is in partial resistance state. When the temperature exceeds the critical temperature, the HTS tape enters normal state.

(3) In the normal state, the relationship between quenching resistance *R* and joule heat *Q* of HTS tape and coil is independent of current waveform. In essence, there is a fixed correspondence between *R* and *Q*. Therefore, *R-Q* curves are consistent under AC and DC overcurrent.

Based on these, the *R-Q* curve method can be used to estimate and predict the quench resistance and temperature change of HTS tapes in completely quenching stage. Compared with FEM, this method can ensure simulation accuracy, greatly shorten calculation time and improve simulation efficiency. It can be used for simulation modeling, prototype design and quench resistance estimation of resistive type SFCL in complex power systems. Furthermore, this method can realize the equivalent calculation of quenching resistance under AC and DC overcurrent, which can provide theoretical support for large-scale DC current limiters to be tested by AC equivalent experiment.

Author Contributions: Conceptualization, S.L. and L.R.; Data curation, J.S.; Formal analysis, S.L. and S.Y.; Funding acquisition, Y.X. and M.S.; Investigation, X.T., Z.L., G.C. and Z.C.; Methodology, S.L. and L.R.; Project administration, T.M. and Y.X.; Resources, T.M. and L.X.; Software, S.L.; Supervision, Y.T.; Validation, X.T., Z.L., G.C., Z.C. and J.S.; Writing—original draft, S.L.; and Writing—review and editing, L.R.

Funding: This research was funded by the Project of the National Natural Science Foundation of China under grant number 51707074 and National Key Research and Development Program 2017YFB0902304.

Conflicts of Interest: The authors declare no conflict of interest.

Nomenclature

BSCCO	Bismuth Strontium Calcium Copper Oxide
YBCO	Yttrium Barium Copper Oxide
1G	First Generation
2G	Second Generation
3D	Three Dimensional
HTS	High Temperature Superconducting
R-SFCL	Resistance type Superconducting Fault Current Limiter
DC	Direct Current
AC	Alternating Current
VSC	Voltage Source Converter
MMC	Modular Multilevel Converter
HVDC	High Voltage Direct Current
FEM	Finite element method
PDE	Partial Differential Equation
RLC	Resistance (R), Inductance (L) and Capacitance (C)

Symbols

U	Voltage (V)		Q	Joule heat (J)
I	Current (A)		P	Power (W)
R	Resistance (Ω)		E	Electric field intensity (V/m)
L	Inductance (H)		J	Current density (A/m^2)
C	Capacitance (F)		r	Resistivity ($\Omega \cdot$m)
t	Time (s)		q_s	Volume power density (W/m)
m	Mass (kg)		k	Heat transfer coefficient (W/(m$^2 \cdot$K))
ρ	Mass density (kg/m^3)		μ	Permeability (H/m)
c	Specific heat capacity (J/(kg·K))		H	Magnetic field intensity (A/m)
q	Conduction heat flux (W/m^2)		B	Magnetic induction intensity (T)
T	Temperature (K)			

Appendix A

The following are the material properties required in the simulation, which mainly come from [46,49,50]. After collection and collation, the properties of materials are shown in Figure A1 and Table A1.

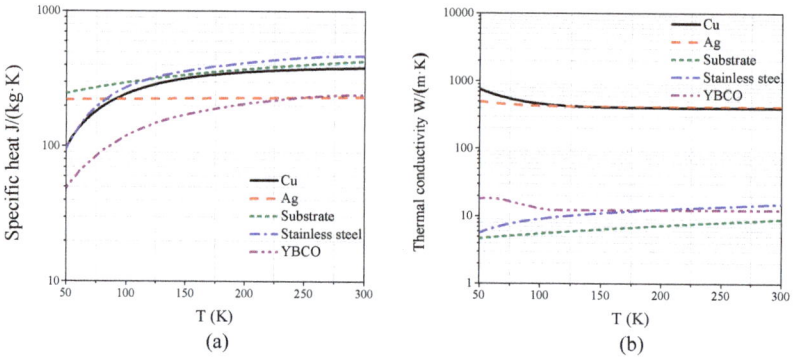

Figure A1. *Cont.*

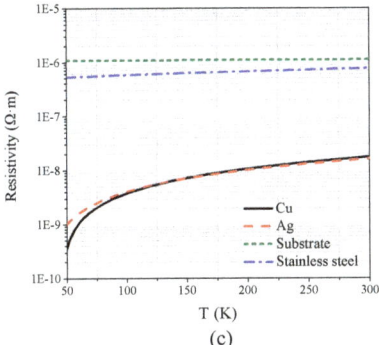

(c)

Figure A1. Material properties: (**a**) specific heat capacity of each layer of material; (**b**) thermal conductivity of each layer of material; and (**c**) resistivity of each layer of material.

Table A1. Mass density of each material in superconducting tape.

	Mass Density (kg/m^3)		
Cu	8940	YBCO	5900
Ag	10500	Stainless steel	7930
Substrate	8910		

References

1. Malozemoff, A.P. Second-Generation High-Temperature Superconductor Wires for the Electric Power Grid. *Annu. Rev. Mater. Res.* **2012**, *42*, 373–397. [CrossRef]
2. Rogalla, H.; Kes, P.H. (Eds.) *100 Years of Superconductivity*; CRC Press: London, UK, 2011.
3. Ronald, M.; Alexis, P.; David, C. Superconducting Materials for Large Scale Applications. *IEEE Trans. Appl. Supercond.* **2004**, *92*, 1639–1654.
4. Malozemoff, A.P.; Verebelyi, D.T. HTS Wire: Status and prospects. *Phys. C. Superconduct.* **2003**, *386*, 424–430. [CrossRef]
5. Chen, Y.; Bian, W.; Huang, W. High critical current density of YBa$_2$Cu$_3$O$_{7-x}$ superconducting films prepared through a DUV-assisted solution deposition process. *Sci. Rep.* **2016**, *6*, 38257. [CrossRef]
6. Schoop, U.; Rupich, M.W.; Thieme, C. Second generation HTS wire based on RABiTS substrates and MOD YBCO. *IEEE Trans. Appl. Supercond.* **2005**, *15*, 2611–2616. [CrossRef]
7. Rupich, M.W.; Schoop, U.; Verebelyi, D.T. The Development of Second Generation HTS Wire at American Superconductor. *IEEE Trans. Appl. Supercond.* **2007**, *17*, 3379–3382. [CrossRef]
8. Pascal, P.T.; Badel, A.; Auran, G. Superconducting Fault Current Limiter for Ship Grid Simulation and Demonstration. *IEEE Trans. Appl. Supercond.* **2017**, *27*, 5601705. [CrossRef]
9. Yang, K.; Yang, Y.; Junaid, M. Direct-Current Vacuum Circuit Breaker with Superconducting Fault-Current Limiter. *IEEE Trans. Appl. Supercond.* **2018**, *28*, 560010. [CrossRef]
10. Chen, L.; Zhang, X.; Qin, Y. Application and Design of a Resistive-Type Superconducting Fault Current Limiter for Efficient Protection of a DC Microgrid. *IEEE Trans. Appl. Supercond.* **2019**, *29*, 5600607. [CrossRef]
11. He, H.; Chen, L.; Yin, T. Application of a SFCL for Fault Ride-Through Capability Enhancement of DG in a Microgrid System and Relay Protection Coordination. *IEEE Trans. Appl. Supercond.* **2016**, *26*, 5603608. [CrossRef]
12. Li, B.; He, J. Studies on the Application of R-SFCL in the VSC-Based DC Distribution System. *IEEE Trans. Appl. Supercond.* **2016**, *26*, 5601005. [CrossRef]
13. Chen, L.; Chen, H.; Shu, Z. Comparison of Inductive and Resistive SFCL to Robustness Improvement of a VSC-HVDC System with Wind Plants against DC Fault. *IEEE Trans. Appl. Supercond.* **2016**, *26*, 5603508. [CrossRef]
14. Xiao, L.; Dai, S.; Lin, L. HTS Power Technology for Future DC Power Grid. *IEEE Trans. Appl. Supercond.* **2013**, *23*, 5401506. [CrossRef]

15. Zhou, Y.; Song, Q.; Guo, F. Quench developing process of HTS tapes under sinusoidal over-currents. *IEEE Trans. Appl. Supercond.* **2005**, *15*, 1651–1654. [CrossRef]
16. Baldan, C.A.; Lamas, J.S.; Shigue, C.Y. Test of a Modular Fault Current Limiter for 220 V Line Using YBCO Coated Conductor Tapes with Shunt Protection. *IEEE Trans. Appl. Supercond.* **2011**, *21*, 1242–1245. [CrossRef]
17. Liu, X.; Wen, J.; Zeng, W. Quenching Characteristics of Different Types of Superconducting Fault Current Limiting Modules. *IEEE Trans. Appl. Supercond.* **2015**, *25*, 5601205. [CrossRef]
18. Zhang, Z.; Yang, J.; Qiu, Q. Research on resistance characteristics of YBCO tape under short-time DC large current impact. *Cryogenics* **2017**, *84*, 53–59. [CrossRef]
19. Xiang, B.; Zhang, L.; Tan, Y. DC current withstanding characteristics of superconductor. In Proceedings of the 3rd International Conference on Electric Power Equipment—Switching Technology, Busan, Korea, 25–28 October 2015; pp. 217–220.
20. Jiang, Z.; Wang, Y.; Dai, S. Influence of Insulation on Quench and Recovery of YBCO Tape under DC Impact. *IEEE Trans. Appl. Supercond.* **2018**, *29*, 7700305. [CrossRef]
21. Xiang, B.; Junaid, M.; Gao, L. Influencing Factors on Quench and Recovery of YBCO Tapes for DC Superconducting Fault Current Limiter. *IEEE Trans. Appl. Supercond.* **2018**, *29*, 5600806. [CrossRef]
22. Xiang, B.; Junaid, M.; Gao, L. Effects of Short Circuit Currents on Quench and Recovery Properties of YBCO Tapes for DC SFCL. *IEEE Trans. Appl. Supercond.* **2018**, *29*, 5600706. [CrossRef]
23. Badel, A.; Antognazza, L.; Decroux, M. Hybrid Model of Quench Propagation in Coated Conductors Applied to Fault Current Limiter Design. *IEEE Trans. Appl. Supercond.* **2013**, *23*, 5603705. [CrossRef]
24. Tong, Y.; Guan, M.; Wang, X. Theoretical estimation of quench occurrence and propagation based on generalized thermoelasticity for LTS/HTS tapes triggered by a spot heater. *Supercond. Sci. Technol.* **2017**, *30*, 045002. [CrossRef]
25. Colangelo, D.; Dutoit, B. Analysis of the influence of the normal zone propagation velocity on the design of resistive fault current limiters. *Supercond. Sci. Technol.* **2014**, *27*, 124005. [CrossRef]
26. Núñez-Chico, A.B.; Martínez, E.; Angurel, L.A.; Navarro, R. Enhanced quench propagation in 2G-HTS coils co-wound with stainless steel or anodised aluminium tapes. *Supercond. Sci. Technol.* **2016**, *29*, 085012. [CrossRef]
27. Hong, Z.; Sheng, J.; Yao, L.; Gu, J.; Jin, Z. The Structure, Performance and Recovery Time of a 10 kV Resistive Type Superconducting Fault Current Limiter. *IEEE Trans. Appl. Supercond.* **2012**, *23*, 5601304. [CrossRef]
28. Ge, H.; Yang, K.; Junaid, M. A quenching recovery time test method for resistive type superconducting fault current limiters used in DC circuit. In Proceedings of the 4th International Conference on Electric Power Equipment—Switching Technology, Xi'an, China, 22–25 October 2017; pp. 393–396.
29. Yang, D.G.; Song, J.B.; Choi, Y.H. Quench and Recovery Characteristics of the Zr-Doped (Gd,Y) BCO Coated Conductor Pancake Coils Insulated With Copper and Kapton Tapes. *IEEE Trans. Appl. Supercond.* **2011**, *21*, 2415–2419. [CrossRef]
30. Lim, S.H.; Lim, S.T. Current Limiting and Recovery Characteristics of a Trigger-Type SFCL Using Double Quench. *IEEE Trans. Appl. Supercond.* **2018**, *28*, 5601305. [CrossRef]
31. Schwarz, M.; Schacherer, C.; Weiss, K.P. Thermodynamic behaviour of a coated conductor for currents above Ic. *Supercond. Sci. Technol.* **2008**, *21*, 054008. [CrossRef]
32. Bagrets, N.; Otten, S.; Weiss, K.P. Thermal and mechanical properties of advanced impregnation materials for HTS cables and coils. *IOP Conf. Ser. Mater. Sci. Eng.* **2015**, *102*, 012021. [CrossRef]
33. Bae, J.H.; Eom, B.Y.; Sim, K.D. Minimum Quench Energy Characteristic of YBCO Coated Conductor with Different Stabilizer Thickness. *IEEE Trans. Appl. Supercond.* **2013**, *23*, 4600404.
34. Falorio, I.; Young, E.A.; Yang, Y. Quench Characteristic and Minimum Quench Energy of 2G YBCO Tapes. *IEEE Trans. Appl. Supercond.* **2015**, *25*, 6605505. [CrossRef]
35. Du, H.I.; Kim, Y.J.; Lee, D.H. Study on Maximum Operating Condition of Resistive Type SFCL Using YBCO Coated Conductor. *IEEE Trans. Appl. Supercond.* **2010**, *20*, 1238–1241.
36. Levin, G.A.; Jones, W.A.; Novak, K.A. The effects of superconductor-stabilizer interfacial resistance on quenching of a pancake coil made out of coated conductor. *Supercond. Sci. Technol.* **2011**, *24*, 035015. [CrossRef]
37. Kwon, N.Y.; Kim, H.S.; Kim, K.L. The Effects of a Stabilizer Thickness of the YBCO Coated Conductor (CC) on the Quench/Recovery Characteristics. *IEEE Trans. Appl. Supercond.* **2010**, *20*, 1246–1249. [CrossRef]

38. Du, H. Evaluation on Resistance Tendency and Recovery Characteristics of 2G Wire with Insulation Layer. *IEEE Trans. Appl. Supercond.* **2013**, *23*, 6602004.
39. The Shanghai Superconductor Website 2019. Available online: http://www.shsctec.com/en/index (accessed on 20 March 2019).
40. Yang, J.; Fletcher, J.; O'Reilly, J. Short-Circuit and Ground Fault Analyses and Location in VSC-Based DC Network Cables. *IEEE Trans. Ind. Electron.* **2012**, *59*, 3827–3837. [CrossRef]
41. Li, C.; Zhao, C.; Xu, J. A Pole-to-Pole Short-Circuit Fault Current Calculation Method for DC Grids. *IEEE Trans. Power Syst.* **2017**, *32*, 4943–4953. [CrossRef]
42. Hong, Z.; Campbell, A.M.; Coombs, T.A. Numerical solution of critical state in superconductivity by finite element software. *Supercond. Sci. Technol.* **2006**, *19*, 1246–1252. [CrossRef]
43. Duron, J.; Grilli, F.; Antognazza, L. Finite-element modelling of YBCO fault current limiter with temperature dependent parameters. *Supercond. Sci. Technol.* **2007**, *20*, 338–344. [CrossRef]
44. Stavrev, S.; Grilli, F.; Dutoit, B. Comparison of numerical methods for modeling of superconductors. *IEEE Trans. Magn.* **2002**, *38*, 849–852. [CrossRef]
45. Curras, S.R.; Vina, J.; Ruibal, M. Normal-state resistivity versus critical current in YBa$_2$Cu$_3$O$_{7-\delta}$ thin films at high current densities. *Phy. C. Superconduct.* **2002**, *372*, 1095–1098. [CrossRef]
46. Roy, F.; Dutoit, B.; Grilli, F. Magneto-Thermal Modeling of Second-Generation HTS for Resistive Fault Current Limiter Design Purposes. *IEEE Trans. Appl. Supercond.* **2008**, *18*, 29–35. [CrossRef]
47. Liang, F.; Yuan, W.; Baldan, C. Modeling and Experiment of the Current Limiting Performance of a Resistive Superconducting Fault Current Limiter in the Experimental System. *J. Supercond. Novel Magn.* **2015**, *28*, 2669–2681. [CrossRef]
48. Frost, W.; Harper, W.L. *Heat Transfer at Low Temperatures*; Plenum: New York, NY, USA, 1975; Chapter 4.
49. De Sousa, W.; Näckel, O.; Noe, M. Transient Simulations of an Air-Coil SFCL. *IEEE Trans. Appl. Supercond.* **2014**, *24*, 5601807. [CrossRef]
50. Kalsi, S.S. *Applications of High Temperature Superconductors to Electric Power Equipment*; John Wiley & Sons: Hoboken, NJ, USA, 2011; pp. 137–138.

 © 2019 by the authors. Licensee MDPI, Basel, Switzerland. This article is an open access article distributed under the terms and conditions of the Creative Commons Attribution (CC BY) license (http://creativecommons.org/licenses/by/4.0/).

Article

How to Choose the Superconducting Material Law for the Modelling of 2G-HTS Coils

Bright Chimezie Robert, Muhammad Umar Fareed and Harold Steven Ruiz *

Department of Engineering and Leicester, Institute for Space & Earth Observation Science, University of Leicester, Leicester LE17RH, UK
* Correspondence: dr.harold.ruiz@leicester.ac.uk

Received: 25 July 2019; Accepted: 19 August 2019; Published: 22 August 2019

Abstract: In an attempt to unveil the impact of the material law selection on the numerical modelling and analysis of the electromagnetic properties of superconducting coils, in this paper we compare the four most common approaches to the E-J power laws that serve as a modelling tool for the conductivity properties of the second generation of high-temperature superconducting (2G-HTS) tapes. The material laws considered are: (i) the celebrated E-J critical-state like-model, with constant critical current density and no dependence with the magnetic field; (ii) the classical Kim's model which introduces an isotropic dependence with the environment magnetic field; (iii) a semi-empirical Kim-like model with an orthonormal field dependence, $J_c(\mathbf{B})$, widely used for the modelling of HTS thin films; and (iv) the experimentally measured E–J material law for SuperPower Inc. 2G-HTS tapes, which account for the magneto-angular anisotropy of the in-field critical current density $J_c(\mathbf{B}; \theta)$, with a derived function similar to Kim's model but taking into account some microstructural parameters, such as the electron mass anisotropy ratio (γ) of the superconducting layer. Particular attention has been given to those physical quantities which within a macroscopic approach can be measured by well-established experimental setups, such as the measurement of the critical current density for each of the turns of the superconducting coil, the resulting distribution of magnetic field, and the curve of hysteretic losses for different amplitudes of an applied alternating transport current at self-field conditions. We demonstrate that although all these superconducting material laws are equally valid from a purely qualitative perspective, the critical state-like model is incapable of predicting the local variation of the critical current density across each of the turns of the superconducting coil, or its non-homogeneous distribution along the width of the superconducting tape. However, depending on the physical quantity of interest and the error tolerance allowed between the numerical predictions and the experimental measurements, in this paper decision criteria are established for different regimes of the applied current, where the suitability of one or another model could be ensured, regardless of whether the actual magneto angular anisotropy properties of the superconducting tape are known.

Keywords: superconducting coil; alternating current (AC) losses; superconducting material law

1. Introduction

In recent years, advances in the development of high-temperature superconducting coils with rare-earth barium-copper oxide (REBCO)-coated conductors have drawn significant attention by the community of researchers in applied superconductivity, thanks to the vast progress in the technology of thin films that has enabled the fabrication of the second generation of high-temperature superconducting (2G-HTS) tapes in the past decade. Thence, the formulation of modelling tools for describing the electromagnetic and thermal properties of such 2G-HTS tapes and their use in high-power-density coils for applications such as superconducting fault current limiters [1–3], transformers [4,5], power generators [6–8], motors [9–12], energy storage systems [13–15], permanent magnets [16,17], and magnetic imaging machines [18,19] is currently a motivating force of intensive research due to the inherent complexity of the material law that governs the electrical properties of the superconducting compound, the computational challenges that are imposed by the large cross-sectional aspect ratio of the 2G-HTS tape, and ultimately, the actual size of the coils that need to be modelled before investing in usually large and customised cryogenic facilities for their experimental testing.

Moreover, multiple publications have already reported that the electrical properties of the large majority of 2G-HTS tapes can be strongly influenced by the intensity and direction of externally and self-induced applied magnetic fields [20–22], although no strong agreement has been reached in terms of the material law that governs their critical current density properties, as widely different material laws can render similar results depending on the physical quantity being studied [22–24]. In this sense, the simulation of the electromagnetic behaviour of 2G-HTS coils presents serious challenges, especially in conditions where one needs to consider the in-field dependence of the critical current, I_c, this in terms of the direction and intensity of the magnetic field per coil-turn, as pointed out in early experiments measuring the magnetic field distribution and AC loss of HTS thin films in superconducting coils [25,26]. Nonetheless, regardless of the 2G-HTS tape being used, certain consensus has been reached in terms of describing the current-voltage characteristics of all type-II superconductors as a power law, $V \propto (I/I_c)^n$, with $n \gg 1$, which from the computational point of view renders the well-known form of the material law for the electric field, also called the E-J power law, $\mathbf{E}(\mathbf{J}) = E_0 \cdot \mathbf{J}/J_c \cdot (|\mathbf{J}|/J_c)^{n-1}$, which in a local but macroscopical approach allows solution of the Maxwell equations inside the superconducting domains within diverse mathematical formulations, for a large range of experimental measurements [16,23,27–38]. Here, J_c is the critical current density of the 2G-HTS tape defined within the standard electric field criterion $E_0 = 1$ µV/cm, which is commonly measured under direct current–voltage measurements with I-V curves typically showing power factors $n > 20$. However, most of the already conducted research has been focused either on the simplification of numerical models to enable faster computational algorithms for relatively large or complex geometries or on the reproduction of specific experimental evidence, such as the hysteresis losses, magnetisation curves, or the measurement of the magnetic field at specific positions that are not necessarily close to the superconducting domains, where changes in the amplitude of the critical current density can be more notorious. In fact, because of the great complexity associated with the experimental measurement of the critical current density at each turn of a superconducting coil, and given the relatively large uncertainty in the reproducibility of the experimental measurements caused by the non-homogeneity of the electric field inside the superconducting tape (both across its length and along its width), computational models are also commonly used to estimate the actual critical current density of the coils. This is done by matching the numerical results of a particular model with any specific physical quantity that can be measured by further experimental methods. However, this does not resolve the problem of the material law selection, but opens some central questions, such as: "Are computational modellers using a sound material law for the superconducting tapes?" and, "Is this material law valid for the study of any other macroscopic electromagnetic quantity?"

Thus, for this Special Issue of MDPI's *Materials*, we present a comprehensive study of the impact of the E-J material law selection on the electromagnetic properties of superconducting racetrack coils, using the H-formulation as a benchmark [27,28] for solving the partial differential equation (PDE) system of Maxwell equations and the four most popular material law models for type-II superconductors, namely: (i) a simplified critical state (CS)-like-model [23,39]; (ii) the classical Kim's model [40,41]; (iii) an empirical Kim-like model with orthonormal field dependence for SuperPower Inc. [42] SCS4050 2G-HTS tapes [33]; and (iv) the generalised form of the critical current density with magneto angular anisotropy measured on the SCS4050 tapes [20]. The paper is organised as follows. In Section 2, we outline the main characteristics of the material law models considered in this study and introduce the geometry and other physical conditions relevant to the solution of this problem. All the aforementioned material laws can be implemented into any finite element formulation for solving the local Maxwell equations (i.e., within the superconducting domains), which obviously will have to render to an univocal solution where the physics richness of the problem does not lie within the mathematical formulation used, but on the invoked material laws. In this sense, a detailed review on how these formulations can be numerically implemented is not within the scope of this paper; instead, all the necessary parameters and conditions that enable the reproducibility of our results are disclosed here. Then, in Section 3, we discuss the main findings from our numerical results, where similarities and differences between the diverse material laws are highlighted in terms of: (i) the analysis of the local distribution of current inside different turns of the superconducting coil; (ii) the intensity of the magnetic field near and over its innermost turn, this being the coil turn that is more prone to being affected by the self- and mutually induced currents when J_c has a strong dependence with the magnetic field; and (iii) the curve of AC losses of the overall system when the REBCO layer of the 2G-HTS tape is modelled as a magnetically isotropic or anisotropic layer. Finally, in Section 4 we summarise the main conclusions of this study, which ultimately are aimed to help the HTS computational modellers in making sensible decisions about how to choose the superconducting material law for 2G-HTS coils, especially when not only a physical quantity needs to be quantified to match with experimental evidence, but in situations where additional consideration must be given to the computing time demanded.

2. Superconducting Material Law Models

In order to compare the impact of different material law models on the numerical modelling of 2G-HTS coils, below we consider a two-dimensional approach based on the celebrated H-formulation, which can be easily implemented in COMSOL Multiphysics [43]. As a benchmark case, we assumed the cross section of a 20-turn racetrack coil wound on a 5 cm midwidth former with a SuperPower Inc. (Schenectady, NY, USA) SCS4050 tape [42], with a 4 mm width and overall thickness of approximately 0.1 mm, with a self-field critical current density (i.e., the critical current density measured in the absence of external magnetic field) measured for long linear sections of $I_{c0} = 114$ A, within the standard 1 µV cm^{-1} electric field criterion at which $E_c = 1 \times 10^{-4}$ Vm^{-1} [20,44]. This tape is composed by a type-II superconducting thin film made of Yttrium barium copper oxide (YBCO), with an aspect ratio of 1 µm × 4 mm, which was then split within our computations into at least 4 × 115 finite elements, enabling a sufficiently high resolution for the discernment of changes in the flux front profiles and the distribution of current density along and across each of the coil turns. This YBCO layer is fabricated by metal organic chemical vapour deposition (MOCVD) over a 0.2 µm buffer of heteroepitaxial layers, deposited by sputtering on a Hastelloy C-276 substrate of 50 µm, and then coated at the top and bottom by a thin Ag layer of ~2 µm, which provide better and more uniform electrical contact with the Cu electro-thermal stabiliser layers of 20 µm thickness each that are deposited at the top and bottom heterostructures of the 2G-HTS tape. Within the numerical scope of this paper, we neglected the thickness of the buffer layer, as all the calculations presented here are at applied currents under the critical threshold I_{c0} and the materials properties are considered to be homogeneous along the entire length of the 2G-HTS tape used, and this is an ansatz that has been already demonstrated to be a reasonably valid approach for reproducing a wide range of experimental evidence [16,28]. However, given the enormous aspect ratio of the YBCO layer and the circumambient layers, including the outer domain that aims to mimic the electric free field condition for solving Faraday's law (the latter usually called by modellers as the "air" domain), each of the Cu, Ag, and substrate layers were split into 4, 2, and 6 finite elements across their respective thickness, times 115 rectangular elements along their width, therefore matching with the mesh of the YBCO layer. In addition, we implemented a mapped triangular mesh of approximately 60,000 distributed finite elements, emulating an "air" domain of about 1200 cm^2 including the 0.2-mm-thick electrically insulating kapton layers between the coil turns. Then, thanks to the 2D axisymmetrical geometry of the racetrack coil, only one side of the coil needs to be modelled, for which a Dirichlet boundary condition ($H_x = 0$) was imposed at the centre axis of the racetrack former. Finally, for the sake of providing sufficient clarity on the physical effects derived, by the use of different material laws describing the E-J superconducting properties, only self-field conditions are considered in this study, with the racetrack coils subject to an alternating current source (AC) which obeys the function $I_{tr} = I_a \sin(\omega t)$.

The power law of the E-J material law in its most generic form,

$$\mathbf{E}(\mathbf{J}) = E_0 \cdot \frac{\mathbf{J}}{J_c} \cdot \left(\frac{|\mathbf{J}|}{J_c}\right)^{n-1}, \tag{1}$$

can be then synthesised within the four most common approaches for the behaviour of the critical current density, J_c, shown in Table 1, and all of them are within the isotropic hypothesis of parallelism between the vectors $\mathbf{E}$ and $\mathbf{J}$, in principle valid for the symmetry of our problem due to the perpendicularity between the vectors B and J. Likewise, for numerical purposes, it would be possible to assume the exponent n to be sufficiently large to have a sharp increment in the electric field when $I > I_{c0}$, which allows neglect of the thermal agitation of vortices and other statistical effects on the conventional magneto-quasi-steady (MQS) approach [39], and therefore emulation of Bean's classical statements [45,46] for the critical-state model [23]. However, in practical cases where finite differences in the homogeneity properties of the superconducting material need to be averaged, as it is in the case for 2G-HTS tapes, and Equation (1) can still be used as long as the parameters I_{c0} and n are experimentally measured. For this reason, we emphasise that the 2G-HTS considered for this study corresponds to a specific product of one of the major superconducting tape manufacturers, SuperPower Inc., whose SCS4050 [42] tape is probably one of the most common tapes used by their customers, and on which these properties have been extensively measured and verified by many research groups, including the magneto-angular anisotropy characteristics of J_c [20].

Table 1. Material law models and related variables considered within the E-J power law formulation for the rare-earth barium-copper oxide (REBCO) material in Equation (1).

Model	Legend	Simplified Description	Material Law $J_c =$	Microstructure Parameters
Critical-State (CS)-Like model [23,39]	C_M	$J = \pm J_c$	J_{c0}	$J_{c0} = 2.85 \times 10^{10}$ A/m^2 [42]
Kim's Model [40]	K_{M1}	$J_c(\mathbf{B})$	$\dfrac{J_{c0}}{\left(1 + \dfrac{|\mathbf{B}|}{B_0}\right)^\alpha}$	$B_0 = 240$ mT $\alpha = 1.5$ [22,42]
Kim-Like model [47]	K_{M2}	$J_c(B_\parallel, B_\perp)$	$\dfrac{J_{c0}}{\left(1 + \dfrac{\sqrt{k^2|B_\parallel|^2 + |B_\perp|^2}}{B_0}\right)^\alpha}$	$B_0 = 42.65$ mT $\alpha = 0.7$ $k = 0.29515$ [33,42,47]
Magneto-Angular Anisotropy Model [20]	R_M	$J_c(\mathbf{B}, \theta)$	$\dfrac{J_{c0}}{\left(1 + \epsilon_\theta \left(\dfrac{|\mathbf{B}|}{B_0}\right)^\beta\right)^\alpha}$	$\epsilon_\theta = \sqrt{\gamma^{-1} \sin^2(\theta) + \cos^2(\theta)}$ $B_0 = 240$ mT, $\alpha = 1.5$ $\beta = 1, \gamma = 5.02$ [20,22,42]

Thus, with $I_{c0} = 114$ A and the exponent $n = 30.5$, the first of the material law models to be implemented was the so-called critical-state-like model, C_M (see Table 1), where the occurrence of current density profiles is related to a magnetic diffusion process that takes place when the local condition for the isotropic critical-state $\mathbf{J} \leq J_{c0}\hat{\mathbf{E}}$ is violated, where $\hat{\mathbf{E}}$ is the unit vector that defines the direction of the electric field $\mathbf{E}$, with $J = J_{c0}$ if $E \neq 0$. Note that within a purely theoretical framework, this macroscopic model can be seen as a general one, as it allows the main physics features derived from the electromagnetic properties of type-II superconductors within Bean's approach to be captured [45,46]. However, in most practical applications of type-II superconductors including the 2G-HTS tapes, it is well known that in the manufacturing of superconducting materials, achieving widely homogeneous electrical properties is something that cannot be assured ad hoc. This is mainly due to the granular properties that these kinds of materials commonly have, and the possibility of having impurities within the structure of the superconductor (i.e., having added materials or alloys with non-superconducting properties during the fabrication of the HTS material). This situation, which is not necessarily unfavourable for the actual use of superconducting materials, is what gave rise to the flux creep concept in the early 1960s, where the occurrence and motion of local profiles of current density predicted by Bean was not only caused by the thermally activated motion of flux structures (vortices) proposed by Anderson [48], but to the local Lorentz force between collective groups of vortex lines whose intensity depends on the microstructure of the HTS material, and which can be averaged within the semi-empirical parameters B_0 and α introduced by Kim [40,41]. These parameters can be obtained from the the aforementioned I-V measurements, either under self-field conditions or externally applied magnetic fields, the first is designated to the classical version of Kim's model, K_{M1} (Table 1), and the latter to Kim-based models where better fitting functions accounting for the experimental measurements have been found. This is the case of the empirical Kim-like model K_{M2} (Table 1) originally introduced by Thakur et al. [47], which has been regularly used for the modelling of 2G-HTS tapes from different manufactures, including the SCS4050 tape made by SuperPower Inc. [33], where $\mathbf{B}_\parallel$ and $\mathbf{B}_\perp$ are the local orthonormal components of the magnetic flux density, being parallel or perpendicular to the wider surface of the 2G-HTS tape. However, in collaboration with the EPEC superconductivity group at the University of Cambridge in the UK, a more general function for the in-field critical current density of 2G-HTS tapes with magneto-angular anisotropy has been recently introduced by Ruiz [20], whose model accounts not only for the flux creep parameters derived by Kim's approach [40] (i.e., the microstructure parameters B_0 and α in Table 1), but extends this function to incorporate Blatter's angular anisotropy factor ϵ_θ, which is a function of the electron mass anisotropy ratio (γ) of the REBCO layer and the angular direction θ of the local magnetic field. Remarkably, this model is capable of reproducing the full magneto-angular anisotropy properties of the critical current density, $J_c(\mathbf{B}, \theta)$, for a broad set of market-available 2G-HTS tapes [20,21], currently manufactured by companies such as SuperPower Inc. [42], American Superconductor [49], Shanghai Superconductor Technology Co., Ltd. [50], and SuperOx [51]. In this sense, in this paper we called this model R_M, which completes the full set of material laws whose range of validity and scope will be critically analysed in the following section.

3. A Crude Analysis of the Material Law Derivatives

By assuming that the superconducting coil former lies on the zx−plane, the 2D axial symmetry commonly assumed for the modelling of long superconducting racetrack coils implies that, the components of the magnetic field inside each of the turns of the superconducting coil must lie on the "xy−plane", $[H_x, H_y]$, orthonormal to the direction of the flux of current density, J_z, and with the vector of electric field pointing towards the same direction, with the E-J power law in Equation (1) governing the electromagnetic properties of the REBCO layer inside the 2G-HTS tape, and all the other "normal" layers obeying Ohm's law, $E_z = \rho J_z$. Notice that Ohm's law is not being constrained to a threshold value of the current density, as it is the case for the E-J power laws shown in Table 1, what represents the main macroscopical difference between "normal" and "superconducting" materials from the electromagnetic point of view, where Maxwell equations are univocal. Thus, Faraday's law, $\nabla \times \mathbf{E} = -\partial_t \mathbf{B}$, is locally solved via the defined electrical field condition for the different materials, whilst simultaneously, Ampère's law, $\nabla \times \mathbf{B} = \mathbf{J}$, is globally solved by the coupling equation of the so-called H-formulation, $\nabla \times (\rho \nabla \times \mathbf{H}) + \mu \partial_t \mathbf{H} = 0$. Note that in order to solve this PDE system, a couple of further ansatz for the material laws need to be taken first, which although are of common practice in applied superconductivity, are still of great importance as they complete the full definition of material laws that are invoked in the electromagnetic modelling of superconducting heterostructures such as 2G-HTS tapes. For instance, the reader might have noticed that within the common use of the H-formulation, a magnetically isotropic and homogeneous behaviour for the magnetic permeability of the different materials is being assumed (μ), such that the magnetic flux density vector $\mathbf{B}$ can be written in terms of the magnetic field vector $\mathbf{H}$, as $\mathbf{B} = \mu \mathbf{H}$, where μ can approach the magnetic permeability of vacuum (μ_0) for all domains in the system to be modelled, as long as the different materials involved behave as non-dispersive (frequency-independent) mediums, with non-magnetic intrinsic properties in their "normal" state. This assumption can be carried out for the REBCO layer in its "superconducting" state, as long as the frequency of the external excitations (i.e., the applied transport current or any external magnetic field) is assumed as constant and it does not change the microstructural properties of the superconducting material [47]. Moreover, as in our definition of Ampère's law, we have invoked the MQS approach [52], where the displacement current densities $\delta_t \mathbf{D}$ are much smaller than $\mathbf{J}$, disappearing in a first-order treatment, it is possible to assume that for uniform and slow sweep rates of the external excitations, the transient variables of electric field $\mathbf{E}$ and electrical resistivity ρ are both small and proportional to $\dot{\mathbf{B}} = \partial B/\partial t$, whereas $\ddot{\mathbf{B}}$, $\dot{\mathbf{E}}$, and $\dot{\rho}$ are negligible. Therefore, defining sufficiently small time steps in the computation of the PDE system, Ampère's law can be rewritten as $\nabla^2 \mathbf{H} - \mu \sigma \partial_t \mathbf{J} = 0$, with approximate integrability condition $\nabla \cdot \mathbf{J} \simeq 0$, where the electrical resistivity function $\rho(J) = \sigma^{-1}$ plays the role of a nonlinear and possibly nonscalar resistivity in the case of the REBCO layer in the superconducting state, that is, $\sigma^{-1} = E_0/|\mathbf{J}| \cdot (|\mathbf{J}|/J_c)^n$, with the critical current density J_c defined in Table 1, and a constant value within the isotropic version of Ohm's law, $\mathbf{J} = \sigma \mathbf{E}$, for all remaining "normal" materials.

Thus, based on the aforementioned physics framework for the material laws of "superconducting" and "normal" materials, and on which the scope of this study must be understood, below we present the main observations derived from the critical analysis on the use of one or another material law, for the modelling of the superconducting properties of the REBCO layer in a 2G-HTS racetrack coil. We pay special attention to measurable (experimental) quantities such as the critical current density, the magnetic field, and the AC losses, all from the local dynamics of the flux of current density across the cross section of the superconducting coil (Figure 1). We encourage the reader to download the high-resolution figures attached to this paper, such that more enhanced visualisation of the figures can be achieved where applicable.

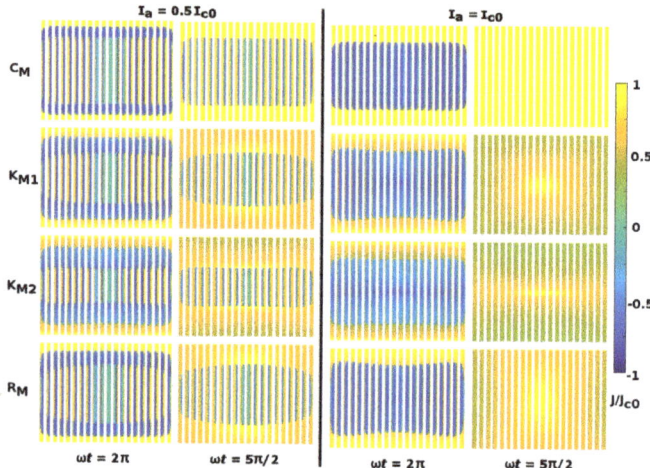

Figure 1. Local profiles of normalized current density J/J_{c0} inside each of the superconducting layers for a 20-turn second generation of high-temperature superconducting (2G-HTS) racetrack coil (not to scale), with the innermost turn being the leftmost 2G-HTS domain (layer) shown in each of the displayed subplots, and with applied AC currents of amplitude $I_a = 0.5\,I_{c0}$ (left pane) and $I_a = I_{c0}$ (right pane), conditioned to the HTS material laws introduced in Table 1. For an easy visualisation of the current-density distributions, results are shown by artificially expanding the thickness of the REBCO layer inside the 2G-HTS tape, as no electrical current flows in any of the other composite layers.

3.1. Dynamics of Flux Front Profiles: A Qualitative Approach

Firstly, we present a qualitative comparison between the four choices for the superconducting material law shown in Table 1, based on which it is possible to analyse how the distribution of current-density profiles change in time for different J_c functions. It is worth mentioning that a full analysis of the time dynamics of flux front profiles inside the HTS domains under the R_M model (excluding the C_M, K_{M1} and K_{M2} models) has already been presented elsewhere [27], where equivalent high meshing considerations were applied in order to obtain a realistic account of the local and global electromagnetic properties of racetrack coils, and where well-defined regions with clear patterns obeying the occurrence of magnetisation currents, transport currents, and flux-free cores can be envisaged for the whole hysteretic behaviour of the racetrack coil with applied transport currents $I_{tr} = I_a \sin(\omega t)$, of amplitudes ranging between $I_a = 0.1 I_{c0}$ and the self-field threshold value $I_a = I_{c0}$. Thus, as the main physical features encountered in the evolution of the flux front profiles for the different material law models have been found to be essentially the same as those reported in [27], for the sake of simplicity, here we display the profiles of current density for only moderate and high applied currents, $I_{tr} = I_a \sin(\omega t)$, with amplitudes $I_a = 0.5 I_{c0}$ and $I_a = I_{c0}$, respectively. These are shown for the two steady steps representing either the self-field condition, $I_{tr}(t) = 0$, or the positive peak of current, $I_{tr}(t) = I_a$, that is, at $\omega t = 2\pi$ and $\omega t = 5\pi/2$, respectively, after completing a full cycle of the external excitation (i.e., where the magnetic relaxation process has already taken place and the electrodynamics of the HTS coil is within a hysteretic behaviour.)

In summary, starting from the C_M model where $J \simeq \pm J_c$, it is possible to observe in a straightforward way, how by analysing the current-density distribution across the thickness of each of the superconducting domains (i.e., within each turn of the 2G-HTS coil), the exact physical nature of the current-density profiles at a local level reveals regions with a concomitant occurrence of positive and negative currents across the thickness of the tape. These both refer explicitly to the self-induced magnetisation currents, which is in good accordance with Bean's model, and they are enclosed by regions within the magneto-transient history of the applied transport current. For instance, for $I_a = 0.5 I_{c0}$ at $\omega t = 2\pi$, the same amount of positive and negative current-density profiles must exist in order to satisfy the global condition $I_{tr} = 0$ and. This implies that starting from the top or bottom edges of the coil towards the centre of each turn, each of the REBCO turns must firstly show a certain amount of current flowing in the positive direction, which then surrounds a weighted region of negative current-density profiles, accounting for the transient history of the applied current, and secondly, a local distribution of the so-called magnetisation currents, whose area decreases as the intensity of the applied current increases, up to its full disappearance at the threshold current condition $I_{tr} = I_{c0}$. In fact, for $I_{tr} < I_{c0}$, the local distribution of current density across the turns of the HTS coil shows a strong dependence on the self-induced magnetisation currents, with a flux-free core visible within the central turns of the superconducting coil. Thus, besides the fact that the intensity of the critical current density for the other material law models in Table 1 cannot be assumed as constant (as is the case for the C_M model), all the above-mentioned physical features can be considered as universal, although slight differences between the shape of the flux front profiles have been observed between the different models, which ultimately will lead to the large quantitative differences observed in the definition of the critical current density at each turn of the HTS coil as well as the measured magnetic field and AC losses, which will be discussed in the following subsections.

In more detail, Figure 1 demonstrates how the local dependence of the critical current density with the magnetic field (included through the material laws K_{M1}, K_{M2}, and R_M) creates not only a reduction in the critical current density across the width of the coil turns, but also a certain deformation of the flux-front profiles envisaged within the C_M model (i.e., the lines of magnetic flux where the slope of J across the width of the tape suddenly changes sign). This is especially evident at the transition between the transport current profiles near the tape edges and the enclosed magnetisation currents whose dynamics will first be explained qualitatively. For instance, starting with the current-density profiles at $\omega t = 2\pi$ (leftmost pane in Figure 1), it can be observed that although the area with negative profiles of transport current (dark-blue profiles) seems to be smaller than the area with positive profiles (light-yellow profiles), the self-field condition $I_{tr} = 0$ is preserved, as the magnitude of J_c decreases from the lateral edges of the REBCO layer towards the centre of each of the coil turns.

Notice that, the imbalance in the critical current density across the width of the REBCO layers (see Figure 2) which is caused by the local variance in the magnetic field (see Figure 3) limits then the area of the magnetisation and flux-free cores of the superconducting coil, with the stronger reduction achieved for the empirical K_{M2} model, where the orthonormal components $B_\parallel$ and $B_\perp$ are explicitly included. However, such a large reduction in the magnetisation core is not observed in the R_M model, whose J_c function was directly derived from experimental measurements in 2G-HTS tapes [20], and where θ defines the angle of attack of the magnetic field vector at a local level (see Table 1). In fact, although the K_{M2} and R_M models are similar in terms of their mathematical structure, given that the conditions $\theta = 0$ and $\theta = \pm\pi/2$ in the R_M model resemble the parallelism and perpendicularity conditions of the magnetic field assumed in the K_{M2} model, the latter substantially increases the influence of the perpendicular component of the magnetic field when arbitrarily lowering the effect of its parallel component through the empirical microstructure parameter k. However, it has already been demonstrated that the actual dependence of J_c with the magnetic field vector for the large majority of 2G-HTS tapes can be accurately described by the semi-analytical model derived by Kim [40,41] for fully isotropic REBCO films on the one hand, and on the other hand, for those materials showing a strong magneto-angular anisotropy [20,21], the local components of the magnetic field in J_c are averaged by the electron mass anisotropy ratio (γ) of the REBCO layer, which at least for the case of the SCS4050 tape, the distribution of flux front profiles at self field conditions between the K_{M1} and R_M models shows nearly the same trend (see Figure 1). This is because strong changes in the critical current density of the SCS4050 tape have been observed only for parallel components of the magnetic field ($\theta = 0, \mathbf{B} = \mathbf{B}_y$) greater than 50 mT [20], which is a condition seen only for about the first three-to-seven innermost and outermost turns of the HTS coil at moderate-to-high applied transport currents, $I_{tr} \subset (0.5I_{c0}, I_{c0})$ (see Figure 3). Then, by observing the full penetration condition (i.e., $I_{tr} = I_{c0}$ at $\omega t = 5\pi/2$ in Figure 1), it is possible to see that with exemption of the C_M model, the critical current density of the superconducting coil changes across the width and thickness of each of the REBCO turns, which is a phenomenon that was experimentally observed in [13,31]. However, in order to be able to properly see how the J_c changes across the different turns of the HTS coil for the different material law models, in the following section we aim to present a more quantitative approach to the above-mentioned observations, which ultimately will be connected to the macroscopic quantities that can be experimentally measured.

3.2. Local Profiles of Current Density: A Quantitative Approach

Continuing with our previous discussion, in Figure 2 we present a more detailed picture of how the measurable critical current density over the surface of the individual turns of the HTS coil not only varies along the width of the superconducting tape, but how from the numerical point of view, these predictions can change as a function of the different material laws that can be invoked.

Beginning our discussion with the C_M model, it has already been mentioned that the critical current density under this model can only have a single value across the domain of the REBCO layers, as can be seen from the dotted lines in Figure 2. This basically defines the threshold value of the critical current density at self-field conditions, J_{c0}, which then has to be lowered by the influence of the magnetic field for magnetically anisotropic superconductors. In this sense, as the intensity of the magnetic field at the innermost and outermost turns of the HTS coil is always greater than the field experienced by the middle turns (see Figure 3), these turns ($T_{th} = 1$ or 20) will always show the lowest critical current density, regardless of the material law $J_c \propto B$. However, as the magnetic field also changes along the width of each of the turns of the HTS coil, with its intensity decreasing from the tape edges towards their geometrical centre; J_c increases from the bottom or top corners of the HTS turns (Figure 1), that is, at $T_w = 0$ and $T_w = 4$ mm in Figure 2, reaching their maximum at the middle distance of the tape width and following the same growing pattern from the innermost and outermost turns towards the centre of the coil, until reaching the magnetic saturation state. This phenomenon can be identified as the plateau in the J_c distribution, at the middle turn ($T_{th} = 10$) for the K_{M1} model (dashed lines) in the peak transport current conditions shown in the right pane plots of Figure 2b,d. Therein, it also can be noticed that with the exception of the innermost and outermost turns of the coil at the peak transport current conditions, the classical K_{M1} model generally overestimates the critical current density across the width of the tape, which is actually one of the main reasons why further approximations for the $J_c\mathbf{B}$ function have been adopted with time [20,22,33,42,47].

On the other hand, when the computations are performed at the threshold condition for the applied transport current, $I_{tr} = I_{c0}$, emulating the actual conditions for the experimental measurement of J_c (see Figure 2d), it can be clearly seen why the empirical and semi-empirical models K_{M2} and R_M, respectively, can render to exactly the same experimental measurements on long sections of individual 2G-HTS tapes. To notice this, it is necessary to bear in mind that in the common practice for the electrical measurement of the I-V curves, and consequently the magnitude of J_c, a series of equally distanced voltage taps are all positioned at the middle width of the HTS tape, where the intensity of the J_c is maximal. Therefore, by observing the local profile of J_c at the central turn of the HTS coil ($T_{th} = 10$ at $T_w = 2$ mm), where the x component of the magnetic field (perpendicular to the surface of the tape) is null, exactly the same intensity of the critical current density can be obtained from the K_{M2} model (dash-dotted yellow line) and the more general R_M model (solid yellow line), although the impact of the parallel component of the magnetic field (B_y) on the local J_c profiles for the HTS coil is underestimated as one moves away from the midpoint. This can also be seen by inspecting the last column of subplots shown in Figure 1, where by colour contrast it can be observed how the maximum J_c allowed by each model is obtained at the midpoint of the coil section, but also how its magnitude decreases towards the coil periphery, where a larger detriment of its magnitude was seen when assuming the K_{M2} model, which is in good agreement with our previous analysis and current numerical results.

Thus, if the physical quantity aimed to be measured and explained in a superconducting coil is the magnitude of the critical current density for the different turns of the 2G-HTS tape being wound, certain caution must be taken in terms of the adequate selection of a material law that describes the complete magneto-angular anisotropic properties of the REBCO layer (from the theoretical perspective), and the correct alignment of the voltage taps in the experimental measurement. Otherwise, differences of up to about 50% between the theoretical and experimental measurements can arise at the innermost and outermost turns of the coil. However, if the critical current density is rather measured by magneto-optical imaging techniques [53–56], where it is possible to observe the full dynamics of the critical current density along the width of the tape surface (T_w), for the self-field and partial penetration conditions of the HTS coil (i.e., at the transport current condition $I_a = 0.5I_{c0}$ and $\omega t = 2\pi$ in Figure 2a), an outstanding resemblance with the experimental measurements reported in [56] was found when the magnetic-field dependence of J_c was considered. In fact, in this paper we confirm that during the initial flux penetration, the actual current distribution inside the YBCO tape does not show any plateau-like feature, as would be expected from the C_M model, but instead it develops strong peaks near the tape edges, as determined by the Kim's hypothesis ($J_c \propto \mathbf{B}$) contained within the K_{M1}, K_{M2}, and R_M models. However, further investigation of the magnetisation properties of single tapes under applied magnetic fields at different orientations (out of the scope of this paper) will need to be conducted to determine if whether the complete validity of one or another model needs to be established beyond the broad success that the R_M model has already achieved over a large set of commercially available 2G-HTS tapes [20,21].

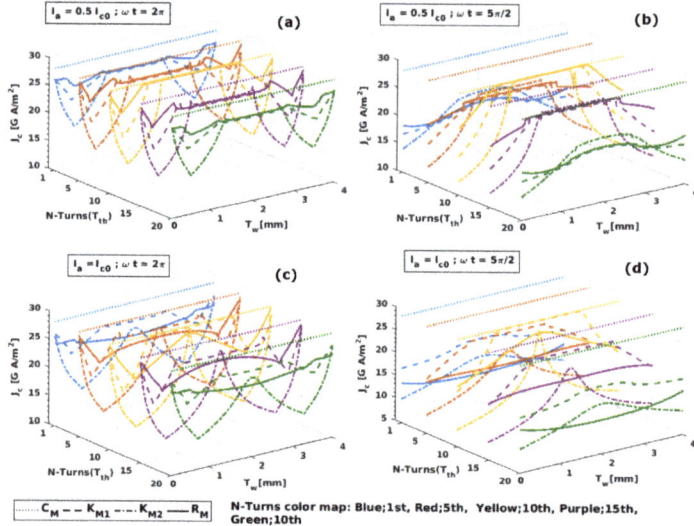

Figure 2. Critical current distribution across the 4-mm-width (T_w) of the 2G-HTS tape measured at the surface of the 1st, 5th, 10th, 15th, and 20th turn (T_{th}) of the modelled coil (Figure 1), under the different material law conditions in Table 1. In this figure we show (**a**) the results for an applied current of amplitude $I_a = 0.5 I_{c0}$ at the self-field condition $\omega t = 2\pi$ and (**b**) the peak transport current condition at $\omega t = 5\pi/2$. Likewise, (**c**) shows the results for an applied current of amplitude equal to the threshold value of the critical current density I_{c0} at $\omega t = 2\pi$ and (**d**) at $\omega t = 5\pi/2$.

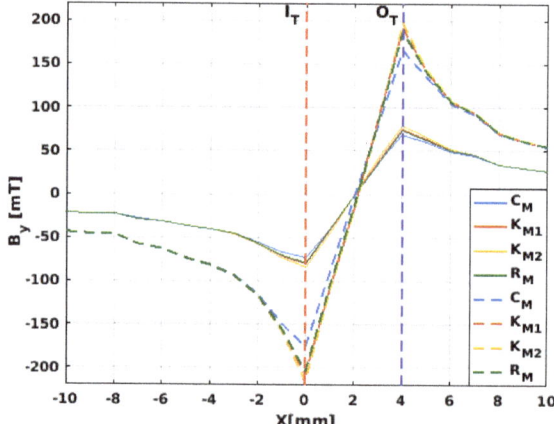

Figure 3. Magnetic field profile over the middle axisymmetric line (x-axis) of the ∼ 4-mm coil section (20 turns) displayed in Figure 1 in the direction parallel to the wider surface of the wound 2G-HTS tape, B_y, measured within and near the coil section for applied currents of amplitude $I_a = I_{c0}$ (dashed curves) and $I_a = 0.5 I_{c0}$ (solid curves), at the peak condition $\omega t = 5\pi/2$. The vertical dashed lines at $x = 0$ mm and $x = \sim 4$ mm respectively refer to the innermost (I_T) and outermost (O_T) turns of the HTS coil.

3.3. Magnetic Field Ratio within Different Material Laws

After having presented a critical discernment of the main physical features that could be observed by the measurement of the local current density in a superconducting coil under the framework of different material laws (Table 1), in this section we present a systematic analysis of how the intensity of the magnetic field component B_y over and near the surface of the innermost turn of the HTS coil (see Figures 4 and 5, respectively) changes as a function of the applied transport current in the self-field condition ($\omega t = 2\pi$), and in the peak transport current condition $I_{tr} = I_a$ at $\omega t = 5\pi/2$.

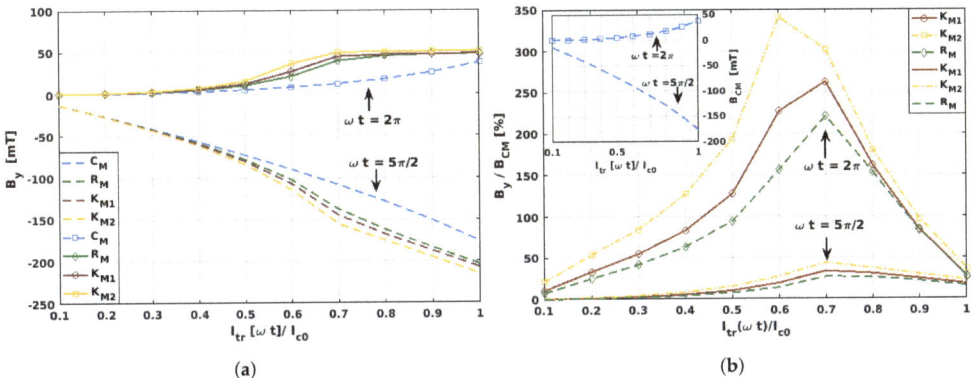

Figure 4. (**a**) Magnetic field component B_y at the middle point of the external surface of the innermost turn of the 2G-HTS coil as a function of the applied transport current $I_{tr}[\omega t]$ at the self-field ($\omega t = 2\pi$) and peak transport current ($\omega t = 5\pi/2$) conditions, and the different material laws presented in Table 1. (**b**) The relative percent ratio between the different material laws, taking as a reference the B_y-field obtained within the critical-state-like-model C_M (inset).

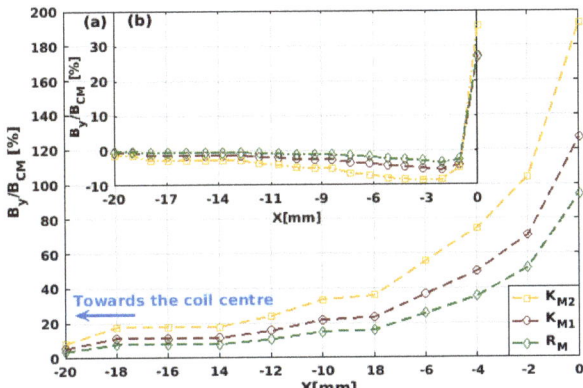

Figure 5. Magnetic field component B_y along the middle axisymmetric line (x-axis) of the 2G-HTS coil calculated within the Kim-based approaches K_{M1}, K_{M2}, and R_M (Table 1) as a percentage function relative to the field predicted by the isotropic C_M model. The measured field is presented from the innermost turn of the coil at 0 mm (see Figure 3) towards the coil centre at the self-field condition $\omega t = 2\pi$, when the applied current has an amplitude of (**a**) $I_a = 0.5 I_{c0}$ (i.e., with the central turns of the HTS coil partly magnetised (see Figure 1)), and (**b**) $I_a = I_{c0}$ (inset), that is, with the HTS coil fully saturated by the transport current and no magnetisation currents.

Note that the apparently arbitrary position on which we decided to display the magnetic field profiles (Figure 4), that is, over the external surface of the innermost coil turn, is indeed the best case for evincing the relative differences between the material laws in Table 1, as negligible changes on the magnetic field were found at the centre of the coil former between all these models (see Figure 5). Thus, if in an actual experimental setup the magnetic field is to be measured at the centre axis of the HTS coil, our numerical results allow us to conclude that the main factor in deciding what material law to use in the computational modellers will not be the dependence of the critical current density with the vector of magnetic field (as in the K_{M1}, K_{M2}, and R_M models), but the extent of the computational time demanded by each of the models. In this sense, it is worth mentioning that, on average, the C_M model always gave the fastest numerical solution, which therefore makes this material law the most suitable model, as long as the interest of the modeller is in targeting the aforementioned experimental conditions.

Evidently, the difference between the magnitude of the magnetic field predicted by the critical-state-like model (C_M) and the Kim-based models (K_{M1}, K_{M2}, and R_M) is larger as the intensity of the applied current increases (see Figure 4). However, through the macroscopic measurement of B_y, a remarkable difference between the C_M and the Kim-based models can be observed as a function of the applied transport current for the self-field ($\omega t = 2\pi$) and peak transport current ($\omega t = 5\pi/2$) conditions, which we found to be intrinsically connected with the dynamics of the flux front profiles described in Section 3.1. In this sense, in Figure 4a it can be seen that for low intensities of the applied current, $I_{tr} \lesssim 0.4I_{c0}$, nearly no difference was obtained in the magnitude of the derived field, meaning that no appreciable difference would be seen between the dimensions of the local flux front profiles between the different material law models. For this reason, the smallest amplitude of the transport current where the local profiles of current density have been shown in this paper is just $I_a = 0.5I_{c0}$ (Figure 1). However, for moderate intensities of the applied current, $\sim 0.4I_{c0} \lesssim I_a \lesssim 0.8I_{c0}$, the magnetic field curves obtained from the Kim-based models rapidly deviate from the isotropic C_M model, all showing a "kink-like" curvature which separates the physical behaviour between what we called moderate and high applied currents. More specifically, the range of high applied currents is characterised by the magnetic saturation of the superconducting coil, and therefore there is a plateau in the measurement of B_y at $\omega t = 2\pi$, meaning that for applied currents $I_{tr} \geq 0.8I_{c0}$, none of the turns of the superconducting coil can exhibit a flux-free core with local condition $J = 0$. On the other hand, the physics of the HTS coil at moderate currents are characterised by transient states, where the Lorentz force between the local profiles of transport current and the induced magnetisation ones is sufficiently strong to reduce the dimensions of the flux-free core by the flux-creep of magnetisation currents, which is added to their own consumption by the extrinsic condition of applied transport current. For a more detailed explanation of this cumbersome phenomenon, which in simple terms corresponds to the actual response of any type-II superconductor to the concomitant action of a magnetic field and a transport current, we encourage the reader to follow references [57–60], where the physical richness of the full dynamics of flux-front profiles in simplified geometries has been exploited.

By adopting a more quantitative approach to illustrate the differences between the use of one or another material law, in Figure 4b we show the percent ratio between the Kim-based material laws and the C_M model (Table 1), where the larger differences were found at the self-field condition $\omega t = 2\pi$. Therein, a clearer picture of the change of the flux creep dynamics between fully isotropic samples which obey the C_M model and those with the magneto-anisotropic properties introduced by the Kim-based models can be seen. For instance, note that in the low-intensity regime ($I_{tr} \lesssim 0.4I_{c0}$), the increment of the B_y/B_{CM} curves primarily obeys a linear function. This means that the physical mechanism leading to the flux creep in the coil turns during the low-current regime is mainly the linear consumption of the magnetisation currents caused by the global condition of transport current, but that at the same time and from local dynamics, it shows that the amount of resulting magnetisation currents inside of each of the coil turns remains nearly the same for all of them. On the other hand, two different phenomena can be recognised for the so-called moderate-intensity regime ($0.4I_{c0} \lesssim I_{tr} \lesssim 0.8I_{c0}$) [27].

Firstly, the field B_y/B_{CM} rapidly increases up to a maximum value, at which the relatively spatial balance between the magnetisation currents and the transport currents (per turn of the coil) cannot be maintained. This leads to a transformation of the flux front profile from a square-like shape towards an elliptical shape (see Figure 1), although during this regime the dimensions of the flux-free core at the centre of the coil are still unaltered. Secondly, when the B_y/B_{CM} curve follows a rapid drop as shown in Figure 4b, the flux creep not only occurs at the positions where there is a consumption of the induced magnetisation currents by the applied transport currents (i.e, where the outer flux front profile can be evinced), but also at the positions where the innermost magnetisation currents are pushed by the superconducting Lorentz force towards the centre of the coil. This reduces the dimensions of the flux-free core, which was previously unaltered. This phenomenon continues until a current level where no evidence of a flux-free core can be observed ($I_{tr} \approx 0.8 I_{c0}$), after which a linear behaviour of B_y/B_{CM} is achieved. Thus, having previous knowledge of the electromagnetic behaviour of an HTS coil within the limits of the C_M model can be considered as a practical approach, not only because this model is more computationally affordable, but because when comparing the actual measurement of the B_y profile near the innermost or outermost turns of the coil with the predicted B_{CM} values, a simple idea of how the local profiles of current density are can be extracted without using more sophisticated experimental techniques. Likewise, in good agreement with our former analysis, for the flux front profiles and the local distribution of current density as a function of the material law, all Kim-based models predict an increment on the intensity of the magnetic field over the surface of the innermost tape of about 2 to 3 times the magnetic field predicted by the C_M model, with the K_M model somehow acting as a mean-field approach for the magneto-angular anisotropic characteristics of REBCO tapes. In this sense, although we demonstrated that the C_M model can be considered as a valid approach when the magnetic field is measured at the centre of the coil former, and also that it can provide some means to elaborate the dynamics of the flux front profiles for magnetically anisotropic materials, we now know that the C_M model cannot be quantitatively accounted if the magnetic field is to be measured near the coil turns. Therefore, instead of the C_M model, the simplified Kim's model K_{M1} can be assumed if a relative tolerance between the experimental and numerical results of ~25% is accepted. Otherwise, and under the expense of possibly increasing the computing time by a factor of 1.5–2, it would be necessary to appeal to a more general approach such as the R_M model.

3.4. AC Losses

Finally, in this section we present our calculated AC loss curves for the benchmarked 20-turn SCS4050 racetrack coil, where the differences between the hysteresis losses that could be predicted by the use of a magnetically isotropic model (e.g., the critical-state-like model, C_M) and the respective Kim-based magneto-anisotropic models R_M, K_{M1}, and K_{M2} (see Table 1) are highlighted (Figure 6). Here, it is worth mentioning that although the C_M model predicts all the macroscopic electromagnetic characteristics of the type-II superconductors well, its validity relies mainly on the qualitative figures of the model, rather than on the quantitative scope added by the Kim-based models. In this sense, in Figure 6a, note how the C_M model generally underestimates the actual AC-losses of the superconducting system, similarly to our previous discussions about the intensity of the magnetic field created by the superconducting coil at different excitation conditions. In fact, the case of a racetrack coil is in good agreement with our previous findings about flux front dynamics and the local distribution of profiles of current density, in which for the low current range ($I_a \leq 0.4 I_{c0}$), all the material law models render to nearly the same hysteretic loss, with relative differences no greater than about twice the losses predicted by the C_M model. This factor might actually seem substantial for some readers, but the AC losses of a type-II superconducting coil can increase by orders of magnitude by just increasing the intensity of the applied current, and in this sense an increment of about two times the estimated losses by the C_M model can be considered, within the ratio of tolerance between the experimental and numerical measurements. However, for moderate-to-high intensities of the applied current, ($I_a \gtrsim 0.4 I_{c0}$), we already established that the impact of the magneto-angular anisotropy of the REBCO layer is very significant. Therefore, the difference with the predicted losses by the C_M model augment significantly as I_a approaches its critical value, I_{c0} (see Figure 6b). For instance, taking as an example this threshold condition for the applied transport current, $I_a/I_{c0} = 1$, the hysteresis losses of the racetrack coil predicted by the C_M model are $L_{CM} = 0.4193$ J/cycle, whilst for the fully magneto-anisotropic model, R_M, the predicted losses are about 4.86 times greater than this value (i.e., $L_{RM} \simeq 2.039$ J/cycle). Likewise, if the empirical J_c-function for the K_{M2} material law is applied, a maximum increment of about 8.5 times the L_{CM} losses are predicted, $L_{KM2} \simeq 3.57$ J/cycle, as a consequence of the increment in the magnitude of the critical current density, caused by the arbitrary reduction of the parallel component of the magnetic field in the function $J_c(B_\|, B_\perp)$, which was assumed by the K_{M2} model (see Table 1). However, once again the most classical and generic approach of the $J_c(B)$ function introduced by Kim, where only spatial but not angular anisotropy is considered (i.e., the K_{M1} model), serves as an average model between the most tailored approaches K_{M2} and R_M, with $L_{KM2} \simeq 2.816$ J/cycle at $I_a = I_{c0}$. Thus, although very strong differences in the predicted energy losses of superconducting racetrack coils could be envisaged by the use of different material laws, especially if the magnetic anisotropic properties of the REBCO layer are considered, we can conclude from this study that if the actual material law governing the J_c properties of the 2G-HTS tape is not known (e.g., the R_M model, where most of the microstructural parameters have already been connected with other physical quantities), then it is advisable to just consider the classical Kim's model for numerical purposes, as long as a tolerance window of about $\pm 28\%$ difference between the experimental and numerical hysteresis losses is allowed.

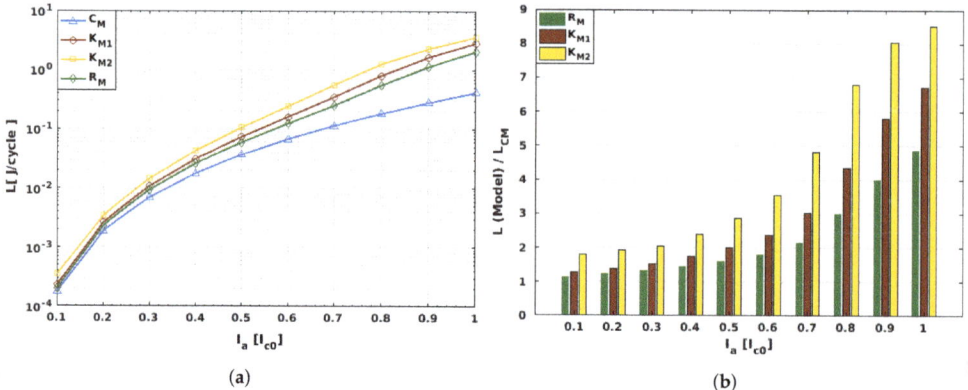

Figure 6. Presented as a function of the amplitude of the applied transport current, I_a, and the material law models introduced in Table 1, in this figure we show (**a**) the predicted AC losses in Joules/cycle for a 20-turn SCS4050 racetrack coil, and (**b**) the relative difference between the Kim-based anisotropic models (R_M, K_{M1}, and K_{M2}) and the magnetically isotropic C_M model.

4. Conclusions

Although many experimental measurements can be at least qualitatively explained by different E-J power law models, for a few decades, some questions about the general validity of the E-J power law remained unsolved. For instance, as the two main criteria for deciding what material law to use in the numerical modelling of a superconducting device are the reproduction of specific experimental evidence (either qualitatively or quantitatively) and the affordability of the computation (due to the usually large demand on memory and processing power that finite element method (FEM)simulations require, especially with the large aspect ratio of geometries such as those implied by the 2G-HTS tapes), it is natural to wonder if the E-J power law invoked by a researcher is a simple artificial function that has been used to match a single piece of experimental evidence. Is the assumed material law model sufficiently valid and well-supported to reproduce all other macroscopical electromagnetic quantities within a more quantitative perspective? Or, in fact, could we in a practical manner use a simpler material law that is capable of accounting for the different electromagnetic phenomena of type-II superconductors, regardless of whether the superconducting material is known to exhibit magneto-anisotropic properties, as is the case for the majority of 2G-HTS tapes? Thus, in an attempt to answer these questions, in this paper we analysed how the selection of different material laws in the numerical modelling of superconducting coils can strongly influence the accurate estimation of macroscopically measurable physical quantities such as the critical current density per coil turn, the magnetic field near the coil armature, and the accounting of the hysteresis losses per cycle, which were all considered under self-field current conditions.

Four different material laws were considered in this study, where a clear impact of the magneto-anisotropic properties of 2G-HTS tapes was disclosed through the direct comparison between an isotropic critical-state-like model, C_M, and different versions of the so-called Kim-based models: K_{M1}, K_{M2}, and R_M in Table 1. These are amongst the most prevalent E-J power law models for 2G-HTS tapes, all validated up to certain extent under different experimental conditions [20,23,33,39,40,47]. We found that although each of these material laws allows a proper qualitative description of the electromagnetism of superconducting coils, substantial quantitative differences were found between their predictions under common experimental conditions, which ultimately, from a purely computational perspective, can help modellers to make a decision on what material law could be more suitable when time and computing power are both limited. In this sense, we concluded that when the physical quantity to be measured is the critical current density turn-by-turn by I-V measurements, certain caution must be taken when compared with the numerical results (Figure 2), as depending on the positioning of the voltage taps, the local magneto-angular anisotropy of the superconducting tape can lead to deviations of up to 50% between the theoretical and experimental measurements. Moreover, if the C_M model is assumed, then no difference in J_c for the different turns of the superconducting coil could be predicted. On the contrary, if a Kim-based model is invoked, then the local variation of the critical current density across the surface of the superconducting tape could be visualised in good agreement with the magneto-optical imaging observations reported in [56]. However, if the intensity of the magnetic field at the central axis of the 2G-HTS coil is the physical quantity of focus (Figure 3), then the simplicity and minimum computing time that can be achieved with the C_M model makes this material law the best option, as no difference was obtained when it was compared with the prognostics of the Kim-based models. Nonetheless, the situation is different if the intensity of the magnetic field near the innermost or outermost turns of the superconducting coil is the desired measurand (Figure 4), as at these locations, local changes in the distribution of current-density profiles can be macroscopically evinced by analysing the B_y behaviour at low, moderate, and high transport currents. In this sense, under this scenario, the simplified Kim's model K_{M1} can be chosen as the most suitable candidate for the numerical modelling of the superconducting properties, as long as a relative tolerance between the experimental and numerical results of ~25% is accepted, and if the expected increase in the computing time cannot be afforded when a more "tailored" approach such as the R_M model is invoked.

Remarkably, when comparing the B_y curve of the Kim-based models with the one derived by the C_M model, over the surface of the innermost turn of the superconducting coil, we obtained an extricable mean for the experimental determination of the magnetic saturation state of the entire coil at self-field conditions and zero transport current ($\omega t = 2\pi$), which is shown in the form of a B_y/B_{CM} plateau for high amplitudes of the applied transport current. Besides, when the study of the superconducting coil focuses on the measurement or estimation of the hysteresis losses, we found that despite the expected underestimation of the AC losses by the C_M model, for low intensities of the applied current ($I_a \leq 0.4 I_{c0}$), all the magneto-anisotropic models led to nearly the same results, with a relative difference of maximum twice the losses expected by the C_M model, which is not necessarily seen to be as large as the hysteretic losses themselves, and can change in orders of magnitude as the intensity of the applied current increases. Nevertheless, for moderate-to-high intensities of the applied current ($I_a \gtrsim 0.4 I_{c0}$), the impact of the magneto-angular anisotropy of the superconducting tape was very significant, with differences found between ~3 to ~8 times the estimated losses of the isotropic C_M model. Evidently, if the computing time and power are not a matter of concern for the modeller of superconducting machines, the best option for choosing the material law for quantitative purposes would then be to select one where the majority of the microstructural parameters for the superconducting tape can or have already been determined by experimental measurements, as is the case of the R_M model. However, if the used 2G-HTS tape is one where the $J_c(\mathbf{B}, \theta)$ is unknown, then we can conclude that the use of the classical Kim's model would be the most advisable choice, if the numerical modellers and experimentalists bear in mind a relative tolerance of ~28% in the estimation of the AC losses.

Author Contributions: B.C.R. contributed to the numerical modelling, simulations, formal analysis, and writing of this paper. B.C.R. and M.U.F. performed post-processing tasks and validation of results. H.S.R. contributed to the supervision, formal analysis, and writing (review and editing) of the paper. All authors have read and approved this manuscript.

Funding: This research was funded by the Engineering and Physical Sciences Research Council, EPSRC grant number EP/S025707/1.

Acknowledgments: All authors acknowledge the support of the East Midlands Energy Research Accelerator (ERA) and, the High Performance Computing Cluster Facility; ALICE at the University of Leicester. B.C.R. thanks the Scholarship unit of the Niger Delta Development Commission for their funding support, and M.U.F. acknowledges the College of Science and Engineering Scholarship Unit of the University of Leicester.

Conflicts of Interest: The authors declare no conflicts of interest.

Abbreviations

The following abbreviations are used in this manuscript:

2D	Two-dimensional
2G-HTS	Second-generation of high-temperature superconductor/superconducting
AC	Alternating current
CS	Critical state
DC	Direct current
FEM	Finite element method
PDE	Partial differential equation
REBCO	Rare-earth barium-copper oxide
YBCO	Yttrium barium copper oxide

References

1. Yan, S.; Ren, L.; Zhang, Y.; Zhu, K.; Su, R.; Xu, Y.; Tang, Y.; Shi, J.; Li, J. AC loss analysis of a flux-coupling type superconducting fault current Limiter. *IEEE Trans. Appl. Supercond.* **2019**, *29*, 1–5. [CrossRef]
2. Zhang, X.; Ruiz, H.; Geng, J.; Coombs, T. Optimal location and minimum number of superconducting fault current limiters for the protection of power grids. *Int. J. Electr. Power Energy Syst.* **2017**, *87*, 136–143. [CrossRef]
3. Ruiz, H.S.; Zhang, X.; Coombs, T.A. Resistive-type superconducting fault current limiters: Concepts, materials, and numerical modeling. *IEEE Trans. Appl. Supercond.* **2015**, *25*, 5601405. [CrossRef]
4. Hellmann, S.; Abplanalp, M.; Elschner, S.; Kudymow, A.; Noe, M. Current limitation experiments on a 1 MVA-class superconducting current limiting transformer. *IEEE Trans. Appl. Supercond.* **2019**, *29*, 1–6. [CrossRef]
5. Pardo, E.; Grilli, F. Numerical simulations of the angular dependence of magnetization AC losses: Coated conductors, Roebel cables and double pancake coils. *Supercond. Sci. Technol.* **2012**, *25*, 014008. [CrossRef]
6. Nam, G.; Sung, H.; Go, B.; Park, M.; Yu, I. Design and comparative analysis of MgB2 and YBCO wire-based-superconducting wind power generators. *IEEE Trans. Appl. Supercond.* **2018**, *28*, 1–5. [CrossRef]
7. Zhong, Z.; Chudy, M.; Ruiz, H.; Zhang, X.; Coombs, T. Critical current studies of a HTS rectangular coil. *Phys. C Supercond. Appl.* **2017**, *536*, 18–25. [CrossRef]

8. Kim, A.; Kim, K.; Park, H.; Kim, G.; Park, T.; Park, M.; Kim, S.; Lee, S.; Ha, H.; Yoon, S.; et al. Performance analysis of a 10-kW superconducting synchronous generator. *IEEE Trans. Appl. Supercond.* **2015**, *25*, 1–4. [CrossRef]
9. Huang, Z.; Ruiz, H.S.; Wang, W.; Jin, Z.; Coombs, T.A. HTS motor performance evaluation by different pulsed field magnetization strategies. *IEEE Trans. Appl. Supercond.* **2017**, *27*, 1–5. [CrossRef]
10. Huang, Z.; Ruiz, H.S.; Zhai, Y.; Geng, J.; Shen, B.; Coombs, T.A. Study of the pulsed field magnetization strategy for the superconducting rotor. *IEEE Trans. Appl. Supercond.* **2016**, *26*, 1–5. [CrossRef]
11. Baghdadi, M.; Ruiz, H.S.; Fagnard, J.F.; Zhang, M.; Wang, W.; Coombs, T.A. Investigation of demagnetization in HTS stacked tapes implemented in electric machines as a result of crossed magnetic field. *IEEE Trans. Appl. Supercond.* **2015**, *25*, 1–4. [CrossRef]
12. Uglietti, D.; Choi, S.; Kiyoshi, T. Design and fabrication of layer-wound YBCO solenoids. *Phys. C Supercond. Appl.* **2010**, *470*, 1749–1751. [CrossRef]
13. Zhu, J.; Chen, P.; Qiu, M.; Liu, C.; Liu, J.; Zhang, H.; Zhang, H.; Ding, K. Experimental investigation of a high temperature superconducting pancake consisted of the REBCO composite cable for superconducting magnetic energy storage system. *IEEE Trans. Appl. Supercond.* **2019**, *29*, 1–4. [CrossRef]
14. Pan, A.V.; MacDonald, L.; Baiej, H.; Cooper, P. Theoretical consideration of superconducting coils for compact superconducting magnetic energy storage systems. *IEEE Trans. Appl. Supercond.* **2016**, *26*, 1–5. [CrossRef]
15. Messina, G.; Morici, L.; Celentano, G.; Marchetti, M.; Corte, A.D. REBCO coils system for axial flux electrical machines application: Manufacturing and testing. *IEEE Trans. Appl. Supercond.* **2016**, *26*, 5205904. [CrossRef]
16. Baghdadi, M.; Ruiz, H.S.; Coombs, T.A. Nature of the low magnetization decay on stacks of second generation superconducting tapes under crossed and rotating magnetic field experiments. *Sci. Rep.* **2018**, *8*, 1342. [CrossRef]
17. Baghdadi, M.; Ruiz, H.S.; Coombs, T.A. Crossed-magnetic-field experiments on stacked second generation superconducting tapes: Reduction of the demagnetization effects. *Appl. Phys. Lett.* **2014**, *104*, 232602. [CrossRef]
18. Moser, E.; Laistler, E.; Schmitt, F.; Kontaxis, G. Ultra-high field NMR and MRI-The role of magnet technology to increase sensitivity and specificity. *Front. Phys.* **2017**, *5*, 33. [CrossRef]
19. Noguchi, S.; Cingoski, V. Simulation of screening current reduction effect in REBCO coils by e+xternal AC magnetic field. *IEEE Trans. Appl. Supercond.* **2017**, *27*, 1–5. [CrossRef]
20. Zhang, X.; Zhong, Z.; Ruiz, H.S.; Geng, J.; Coombs, T.A. General approach for the determination of the magneto-angular dependence of the critical current of YBCO coated conductors. *Supercond. Sci. Technol.* **2017**, *30*, 025010. [CrossRef]
21. Zhang, X.; Zhong, Z.; Geng, J.; Shen, B.; Ma, J.; Li, C.; Zhang, H.; Dong, Q.; Coombs, T.A. Study of critical current and n-values of 2G HTS tapes: Their magnetic field-angular dependence. *J. Supercond. Novel Magn.* **2018**, *31*, 3847–3854. [CrossRef]
22. Wimbush, S.C.; Strickland, N.M. A public database of high-temperature superconductor critical current data. *IEEE Trans. Appl. Supercond.* **2017**, *27*, 1–5. [CrossRef]
23. Badía-Majós, A.; López, C. Modelling current voltage characteristics of practical superconductors. *Supercond. Sci. Technol.* **2015**, *28*, 024003. [CrossRef]
24. Sirois, F.; Grilli, F. Potential and limits of numerical modelling for supporting the development of HTS devices. *Supercond. Sci. Technol.* **2015**, *28*, 043002. [CrossRef]
25. Higashikawa, K.; Nakamura, T.; Hoshino, T. Anisotropic distributions of current density and electric field in Bi-2223/Ag coil with consideration of multifilamentary structure. *Phys. C Supercond. Appl.* **2005**, *419*, 129–140. [CrossRef]
26. Polak, M.; Demencik, E.; Jansak, L.; Mozola, P.; Aized, D.; Thieme, C.L.H.; Levin, G.A.; Barnes, P.N. AC losses in a $YBa_2Cu_3O_{7-x}$ coil. *Appl. Phys. Lett.* **2006**, *88*, 232501. [CrossRef]
27. Robert, B.C.; Fareed, M.U.; Ruiz, H.S. Local electromagnetic properties and hysteresis losses in uniformly and non-uniformly wound 2G-HTS racetrack coils. *arXiv* **2019**, arXiv:1907.09893.
28. Quéval, L.; Zermeño, V.M.R.; Grilli, F. Numerical models for ac loss calculation in large-scale applications of HTS coated conductors. *Supercond. Sci. Technol.* **2016**, *29*, 024007. [CrossRef]
29. Zhang, H.; Zhang, M.; Yuan, W. An efficient 3D finite element method model based on the T-A formulation for superconducting coated conductors. *Supercond. Sci. Technol.* **2017**, *30*, 024005. [CrossRef]

30. Liang, F.; Venuturumilli, S.; Zhang, H.; Zhang, M.; Kvitkovic, J.; Pamidi, S.; Wang, Y.; Yuan, W. A finite element model for simulating second generation high temperature superconducting coils/stacks with large number of turns. *J. Appl. Phys.* **2017**, *122*, 043903. [CrossRef]
31. Martins, F.G.R.; Sass, F.; Barusco, P.; Ferreira, A.C.; de Andrade, R. Using the integral equations method to model a 2G racetrack coil with anisotropic critical current dependence. *Supercond. Sci. Technol.* **2017**, *30*, 115009. [CrossRef]
32. Pardo, E. Calculation of AC loss in coated conductor coils with a large number of turns. *Supercond. Sci. Technol.* **2013**, *26*, 105017. [CrossRef]
33. Zermeno, V.M.R.; Abrahamsen, A.B.; Mijatovic, N.; Jensen, B.B.; Sørensen, M.P. Calculation of alternating current losses in stacks and coils made of second generation high temperature superconducting tapes for large scale applications. *J. Appl. Phys.* **2013**, *114*, 173901. [CrossRef]
34. Prigozhin, L.; Sokolovsky, V. Computing AC losses in stacks of high-temperature superconducting tapes. *Supercond. Sci. Technol.* **2011**, *24*, 075012. [CrossRef]
35. Brambilla, R.; Grilli, F.; Nguyen, D.N.; Martini, L.; Sirois, F. AC losses in thin superconductors: The integral equation method applied to stacks and windings. *Supercond. Sci. Technol.* **2009**, *22*, 075018. [CrossRef]
36. Clem, J.R.; Claassen, J.H.; Mawatari, Y. AC losses in a finite Z stack using an anisotropic homogeneous-medium approximation. *Supercond. Sci. Technol.* **2007**, *20*, 1130. [CrossRef]
37. Ichiki, Y.; Ohsaki, H. Numerical analysis of AC loss characteristics of YBCO coated conductors arranged in parallel. *IEEE Trans. Appl. Supercond.* **2005**, *15*, 2851–2854. [CrossRef]
38. Blatter, G.; Feigel'man, M.V.; Geshkenbein, V.B.; Larkin, A.I.; Vinokur, V.M. Vortices in high-temperature superconductors. *Rev. Mod. Phys.* **1994**, *66*, 1125–1388. [CrossRef]
39. Badía-Majós, A.; López, C.; Ruiz, H.S. General critical states in type-II superconductors. *Phys. Rev. B* **2009**, *80*, 144509. [CrossRef]
40. Kim, Y.B.; Hempstead, C.F.; Strnad, A.R. Critical persistent currents in hard superconductors. *Phys. Rev. Lett.* **1962**, *9*, 306–309. [CrossRef]
41. Kim, Y.B.; Hempstead, C.F.; Strnad, A.R. Resistive states of hard superconductors. *Rev. Mod. Phys.* **1964**, *36*, 43–45. [CrossRef]
42. SuperPower ®2G HTS Wire. Available online: http://www.superpower-inc.com/content/2g-hts-wire (accessed on 14 September 2017).
43. Brambilla, R.; Grilli, F.; Martini, L. Development of an edge-element model for AC loss computation of high-temperature superconductors. *Supercond. Sci. Technol.* **2007**, *20*, 16. [CrossRef]
44. Song, H.; Brownsey, P.; Zhang, Y.; Waterman, J.; Fukushima, T.; Hazelton, D. 2G HTS coil technology development at SuperPower. *IEEE Trans. Appl. Supercond.* **2013**, *23*, 4600806. [CrossRef]
45. Bean, C.P. Magnetization of hard superconductors. *Phys. Rev. Lett.* **1962**, *8*, 250–253. [CrossRef]
46. Bean, C.P. Magnetization of high-field superconductors. *Rev. Mod. Phys.* **1964**, *36*, 31–39. [CrossRef]
47. Thakur, K.P.; Raj, A.; Brandt, E.H.; Kvitkovic, J.; Pamidi, S.V. Frequency-dependent critical current and transport ac loss of superconductor strip and Roebel cable. *Supercond. Sci. Technol.* **2011**, *24*, 065024. [CrossRef]
48. Anderson, P.W. Theory of flux creep in hard superconductors. *Phys. Rev. Lett.* **1962**, *9*, 309–311. [CrossRef]
49. American Superconductor. AMSC Amperium ®HTS Wire. Available online: www.amsc.com/solutions-products/hts_wire.html (accessed on 4 November 2017).
50. Shangai Superconductor Technology Co. Ltd. 2G HTS Strip. Available online: www.amsc.com/solutions-products/hts_wire.html (accessed on 4 November 2017).
51. SuperOx 2G HTS Wire. Available online: http://www.superox.ru/en/products/ (accessed on 4 November 2017).
52. Ruiz, H.S.; Badía-Majós, A. Smooth double critical state theory for type-II superconductors. *Supercond. Sci. Technol.* **2010**, *23*, 105007. [CrossRef]
53. Baziljevich, M.; Johansen, T.H.; Bratsberg, H.; Shen, Y.; Vase, P. Magneto-optic observation of anomalous Meissner current flow in superconducting thin films with slits. *Appl. Phys. Lett.* **1996**, *69*, 3590–3592. [CrossRef]
54. Jooss, C.; Guth, K.; Born, V.; Albrecht, J. Electric field distribution at low-angle grain boundaries in high-temperature superconductors. *Phys. Rev. B* **2001**, *65*, 014505. [CrossRef]

55. Jooss, C.; Albrecht, J.; Kuhn, H.; Leonhardt, S.; Kronmüller, H. Magneto-optical studies of current distributions in high-Tc superconductors. *Rep. Prog. Phys.* **2002**, *65*, 651–788. [CrossRef]
56. Wells, F.S.; Pan, A.V.; Golovchanskiy, I.A.; Fedoseev, S.A.; Rozenfeld, A. Observation of transient overcritical currents in YBCO thin films using high-speed magneto-optical imaging and dynamic current mapping. *Sci. Rep.* **2017**, *7*, 40235. [CrossRef] [PubMed]
57. Ruiz, H.S.; Badía-Majós, A.; López, C. Material laws and related uncommon phenomena in the electromagnetic response of type-II superconductors in longitudinal geometry. *Supercond. Sci. Technol.* **2011**, *24*, 115005. [CrossRef]
58. Ruiz, H.S.; López, C.; Badía-Majós, A. Inversion mechanism for the transport current in type-II superconductors. *Phys. Rev. B* **2011**, *83*, 014506. [CrossRef]
59. Ruiz, H.S.; Badía-Majós, A.; Genenko, Y.A.; Rauh, H.; Yampolskii, S.V. Superconducting wire subject to synchronous oscillating excitations: Power dissipation, magnetic response, and low-pass filtering. *Appl. Phys. Lett.* **2012**, *100*, 112602. [CrossRef]
60. Ruiz, H.S.; Badía-Majós, A. Exotic magnetic response of superconducting wires subject to synchronous and asynchronous oscillating excitations. *J. Appl. Phys.* **2013**, *113*, 193906. [CrossRef]

© 2019 by the authors. Licensee MDPI, Basel, Switzerland. This article is an open access article distributed under the terms and conditions of the Creative Commons Attribution (CC BY) license (http://creativecommons.org/licenses/by/4.0/).

Article

Numerical Study on Transient State of Inductive Fault Current Limiter Based on Field-Circuit Coupling Method

Wenrong Li [1], Jie Sheng [1,*], Derong Qiu [1], Junbo Cheng [2], Haosheng Ye [1] and Zhiyong Hong [1]

1. School of Electronic Information and Electrical Engineering, Shanghai Jiao Tong University, Shanghai 200240, China
2. Russian Representative Office of State Grid Corporation of China, Moscow 109807, Russian
* Correspondence: sjl@sjtu.edu.cn

Received: 13 July 2019; Accepted: 28 August 2019; Published: 31 August 2019

Abstract: As the capacity of the power grid continues to expand, high-level fault currents might be caused during a contingency, and the problem of short-circuit current over-limitation is imminent. The high-temperature superconducting (HTS) fault current limiter (FCL) is an effective method to solve this problem. In this paper, a transient numerical model for the process of limiting current in the inductive FCL is proposed. The model is based on the coupling of multiphysics finite element simulation and a circuit model. The voltage source is used as input, which can simulate the macroscopic characteristics in the process of limiting current, such as the voltage and current waveforms, and can also simulate microscopic characteristics, such as temperature, magnetic field, and electrodynamic force distribution. The short-circuit experimental data of an air core inductive superconducting fault current limiter (SFCL) prototype was compared with the simulation results to verify the reliability of the simulation.

Keywords: inductive fault current limiter; magnetic flux shielding; multiphysics simulation; transient state; field-circuit coupling method

1. Introduction

With the increasing power load and increased short-circuit capacity, the short-circuit current of the grid will continue to rise and will gradually approach the limits of the breaking capacity, which will greatly affect the stability of the power system and equipment. Among the current-limiting equipment, the superconducting fault current limiter (SFCL) is considered as one of the most effective methods to solve the problem of exceeding short-circuit current. Thus, simulating and conducting experimental research on it is significant [1].

SFCLs can be divided into resistive, inductive, three-phase reactor, saturated iron core, and bridge types. The principle of resistive SFCL is the transition of the superconductor from superconducting to normal state. They are relatively simple in structure, but continuous losses during rated operation is unavoidable. The inductive SFCL omit the current leads, but in general use an iron core, which makes this SFCL relatively heavy and costly. Recently, the inductive SFCL has received more attention [2–5].

In 1996, the Swiss ABB technology company developed a 1.2 MVA/10.5 kV three-phase inductive high-temperature superconducting fault current limiter (HTS FCL), installed it in a hydropower station in Lausanne, and carried out a durability test for one year. It was the first superconducting electrical equipment tested under the actual operating conditions of a power station [6]. The secondary winding is a superconducting magnetic shielding ring made of BSCCO-2212, cooled by liquid nitrogen. In the test, the peak value of short-circuit current was limited to a range from 60 kA to 700 A. The Warsaw Institute of Superconductivity in Poland and the Karlsruhe Institute of Technology in Germany developed an air-core inductive FCL. In 2016 and 2018, 15 kV/140 A [7] and 10 kV/600 A [8] single-phase inductive FCLs were developed.

The inductive current limiter simulation is divided into a steady-state simulation for rated operation and a transient-state simulation for the current-limiting process. When the SFCL is in rated operation, the temperature variation of superconducting winding can be neglected, because the HTS tapes do not generated large heat quickly. The steady-state simulation can be based on the numerical model using the H formulation [9,10] or the T-A formulation [11,12]. These numerical models usually use the finite element method (FEM) or the finite difference method to solve Maxwell equations (including Gauss's law for magnetism, Faraday's law of induction, and Ampere's law with Maxwell's addition); the detailed derivation process is shown in the above references. The input of the simulation model is current, and the output is the current density distribution, magnetic field distribution, and AC loss in the SFCL. When the SFCL is in the current-limiting process, the temperature of the superconducting winding rises significantly, affected by the short-circuit current. Since impedance of the superconducting winding is affected by current, magnetic field, and temperature simultaneously, multiple factors are combined, so a transient simulation of the inductive FCL is more difficult.

In [13], resistance connected in parallel with a switch was used to simulate the resistive component of the superconducting winding, and the secondary induced current was calculated by the finite element simulation based on an electromagnetic model. The advantage of the model is that it provided fast computation speed and could simulate the macroscopic working characteristics of the current limiter, including voltage and current curves. However, since the model, lumped parameter model, equated the resistive component of tape to the parallel connection of the resistor and the switch, only the equivalent parameter value can be obtained. It was impossible to simulate microscopic working characteristics, such as the distribution of current and temperature. In [14–16], the magnetic field shielding characteristics of a single inductive current limiter unit in air or iron core were measured at different currents, and then equivalent inductance of the entire current limiter was estimated from the angle of the alternating magnetic flux and the winding hinge. The calculation result was used in the circuit model calculation. Since the model combined the measured values, the calculation results were more accurate. However, the measured value of a single current-limiting unit cannot fully represent the electromagnetic environment of the current limiter. At the same time, since the current-limiter impedance was calculated by the overall equivalent method, it was impossible to simulate the microscopic properties of the tape.

In this paper, a transient numerical model for the process of limiting current in the inductive FCL is proposed. This numerical model including three main models: Circuit model, electromagnetic model, and heat transfer model. Then, an optimization method to accelerate convergence velocity of iteration is proposed. Based on the model, several typical examples are discussed, including the following parameters: The current and voltage, magnetic field and electrodynamic force distribution, and temperature distribution. A SFCL prototype is fabricated, and a short-circuit test of the SFCL is compared with the simulation results to verify the reliability of the simulation.

2. Methodology of Numerical Model

2.1. Model Overview

When a short-circuit fault occurs in the power grid, the short-circuit current will far exceed the critical current in the current limiter. Meanwhile, part of the current will flow through the metal layer, and a large amount of joule heat will cause a significant temperature rise in the winding. Therefore, to calculate the temperature change inside the winding, the temperature field in a simulation model is needed, which is combined with the electromagnetic model to calculate the current-limiter impedance. In addition, in actual situations, the voltage value of the transformer is known instead of the short-circuit current, so the circuit model is used to calculate the instantaneous short-circuit current based on the power voltage and current-limiter impedance. Since the calculation results of multiple physical fields involved in the whole model are interdependent (such as current and current-limiter impedance), multilayer iterative calculations are needed to converge the model and simulate the operating characteristics of the actual current limiter [17,18].

The sophisticated commercial FEM software can solve the complex problems quickly and accurately. The ability of interactive post-processing and visualization make the analyzing of the results easier. The FEM software being used in this paper is COMSOL multiphysics 3.0, and both electromagnetic model and heat transfer model are based on the software. The circuit model is calculated by MATLAB.

The overall calculation structure of the model is shown in Figure 1.

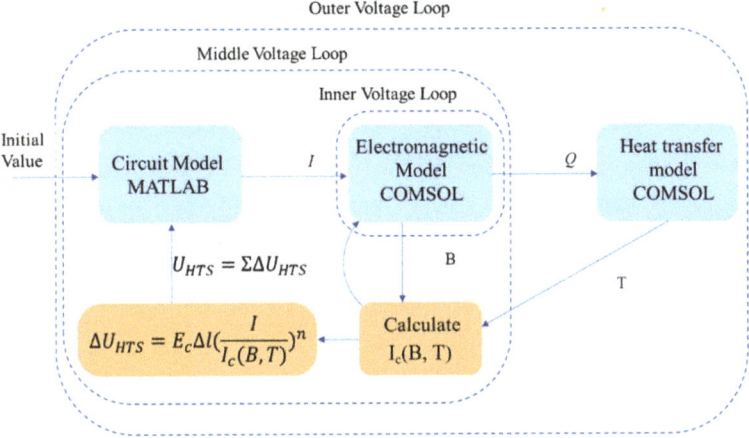

Figure 1. Schematic diagram of inductive superconducting fault current limiter (SFCL) transient-state simulation model.

2.2. Circuit Model

The SFCL model can be equivalent to a transformer with a secondary short circuit. The circuit model of the SFCL is shown in Figure 2.

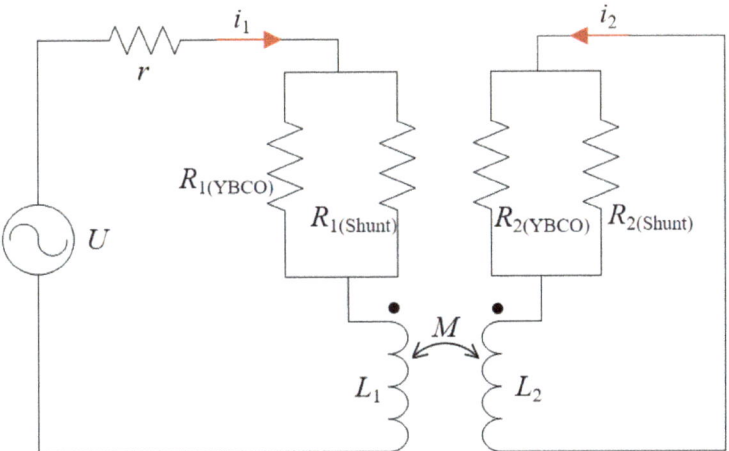

Figure 2. Schematic diagram of circuit model.

In Figure 2, U is the power voltage, r is the line resistance, i_1 and i_2 are the primary and secondary currents, respectively, $R_{1(YBCO)}$ and $R_{1(Shunt)}$ are the equivalent resistance of the HTS layer and metal layer in primary winding, respectively, $R_{2(YBCO)}$ and $R_{2(Shunt)}$ are the same as above in the secondary winding, L_1 and L_2 are self-inductance of the primary and secondary winding, respectively, and M is mutual inductance.

In this model, the calculation of resistance is equivalent to the calculation of the resistance voltage drop. The resistance of the primary and secondary windings is in parallel with the resistance of the HTS layer and the metal layer.

For the HTS layer, the resistive voltage drop is related to the temperature, amplitude, and direction of the magnetic field. Because of the nonuniform magnetic field and different positions of the primary and secondary windings, it is necessary to separately calculate the resistive voltage drop of each turn, and then accumulate the entire winding to obtain the total resistive voltage drop, as shown in Equation (1):

$$U = \sum_N E_c \Delta l \left(\frac{I_{YBCO}}{I_c(B,T)} \right)^n \tag{1}$$

where N is the total number of turns, Δl is the length of each turn, E_c is the critical electrical field, usually taking a constant value 1×10^{-6} V/m, n value is 29 for the tape used in this study. I_{YBCO} is the current of the HTS layer, and $I_c(B,T)$ is the critical current under magnetic field B and temperature T, as shown in Equation (2) [19]:

$$I_c(B,T) = \frac{I_c(T)}{\left[1 + \sqrt{(kB_\parallel)^2 + B_\perp^2}/B_c \right]^\alpha} \tag{2}$$

In the formula, $B_\parallel$ is the components of the magnetic field parallel to the superconducting surface, $B_\perp$ is the components of the magnetic field vertical to the superconducting surface, k, α, B_c are physical parameter of superconducting tape, which represent the degradation of critical under different background magnetic field, and the value of k is between 0 and 1. In this paper, according to the test results of the critical current magnetic field angle-dependency of the $YBa_2Cu_3O_{7-x}$ (YBCO) tape, $k = 0.3672$, $\alpha = 0.6267$, and $B_c = 0.05249$ T.

For the metal layer, resistivity is only related to temperature, but since it is composed of various metals, the resistivity of each metal varies with temperature. The resistivity of superconducting tape as a function of temperature is measured when it is above the critical temperature. Below the critical temperature, a linear cubic spline fit is made at 92 K based on the measured data.

In summary, the overall circuit model formula can be obtained, and discrete iterations are used in the actual calculation, as shown in Equation (3):

$$\begin{cases} U_{(m)} = i_{1(m)}r + \sum_{N_1} E_c \Delta l \left(\frac{i_{1(YBCO)(m)}}{I_c(B,T)} \right)^n + L_1 \frac{i_{1(m)} - i_{1(m-1)}}{\Delta t} - M \frac{i_{2(m)} - i_{2(m-1)}}{\Delta t} \\ 0 = \sum_{N_2} E_c \Delta l \left(\frac{i_{2(YBCO)(m)}}{I_c(B,T)} \right)^n + L_2 \frac{i_{2(m)} - i_{2(m-1)}}{\Delta t} - M \frac{i_{1(m)} - i_{1(m-1)}}{\Delta t} \end{cases} \tag{3}$$

In the formula, $i_{1(YBCO)(m)}$ is taken as an example, the number 1 represents the primary side, (m) represents the mth step iteration, and $(YBCO)$ represents the HTS layer, so $i_{1(YBCO)(m)}$ represents the current of the HTS layer in the primary winding at the mth iteration; $i_{2(m-1)}$ represents the total current in the secondary winding at the $(m-1)$th iteration; N_1 and N_2 is the total number of primary winding and secondary winding, respectively, and Δt is the size of the iterative step (1×10^{-5} s).

2.3. Electromagnetic Model

The electromagnetic model is based on the COMSOL AC/DC module, and it is based on Maxwell equations, as shown in Equation (4):

$$\begin{cases} \nabla \cdot B = 0 \\ J = \nabla \times H \\ \nabla \times E = -\frac{\partial B}{\partial t} \end{cases} \quad (4)$$

where B is the magnetic flux density, J is the current density, H is the magnetic field strength, E is the electric field strength, and t represent time.

Figure 3 shows a schematic diagram of the electromagnetic model, which is a two-dimensional axisymmetric graph. Figure 3b is a diagram with mesh generation in simulation, the thickness of the superconducting layer is magnified 10 times to accelerate the simulation. About mesh generation in superconducting layer, 20 grids in z direction and 1 grid in r direction are generated.

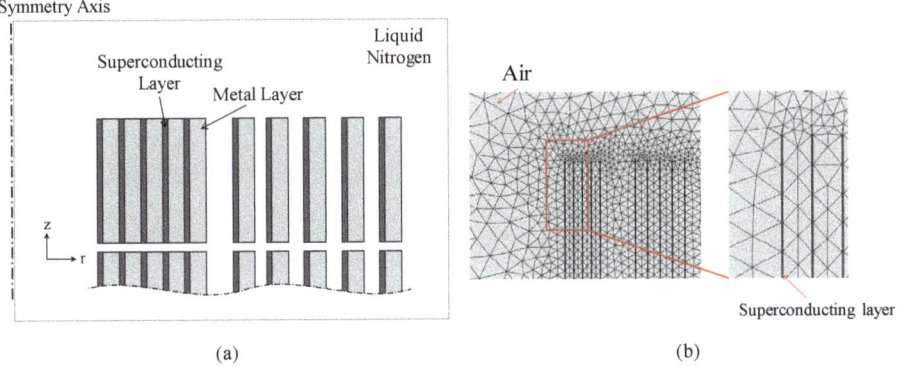

Figure 3. Schematic diagram of electromagnetic model. (**a**) Structure; (**b**) mesh.

In order to simplify the model and speed up the calculation, $B_{\parallel}$ (Equation (2)) is the average of parallel magnetic field in one turn, and $B_{\perp}$ (Equation (2)) is the average of vertical magnetic field in one turn. Differences in current density at different points are ignored for the same turn. The electromagnetic model is used to calculate the I_{YBCO} and I_{shunt} when the total current is determined.

The specifications of the SFCL model are shown in Table 1.

Table 1. Specifications of SFCL model.

Item	Specifications	Value
YBCO tape	Tape width Thickness without insulation Ic/n-value (@77 K, self-field) Critical temperature	4.8 mm 0.25 mm 40 A/29 92 K
Primary winding (insulation)	Number of turns Inner/outer diameter Turn-to-turn gap	40 114 mm/116.8 mm 0.35 mm
Secondary winding (no-insulation)	Number of turns Inner/outer diameter Turn-to-turn gap	40 110 mm/112 mm 0.25 mm

2.4. Heat Transfer Model

The heat transfer model is based on the COMSOL heat transfer in solids module. As shown in Figure 4, a heat flux is applied to the outer boundary to simulate the process of boiling liquid nitrogen heat transfer, so that the simulated liquid nitrogen region can be directly deleted.

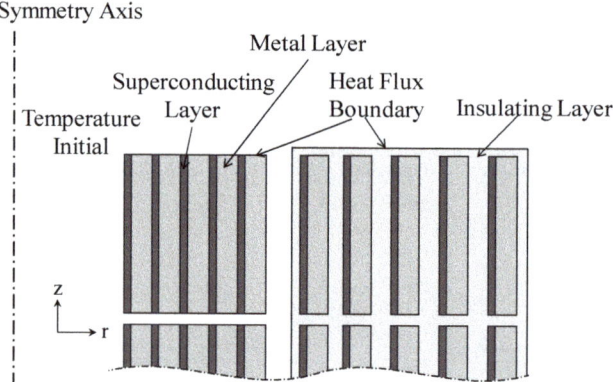

Figure 4. Schematic diagram of heat transfer model.

The heat transfer model is divided into two parts. The first part is solid heat transfer, including heat transfer inside the superconducting tape, between tapes, and between tape and insulating material. The basic equation is as follows [20]:

$$\rho C_p(T) \frac{\partial T}{\partial t} = \nabla \cdot (\lambda(T) \nabla T) + Q \quad (5)$$

where ρ is density, T is temperature, $C_p(T)$ is specific heat capacity, $\lambda(T)$ is thermal conductivity, t represent time, and Q is the heat source in this model, which is the density loss $E \cdot J$ of the winding.

In the model, the simulation space is divided into three parts: Superconducting, metal, and insulating layers. The metal layer is composed of multiple layers of different materials, and the specific heat capacity $C_p(T)$ and thermal conductivity $\lambda(T)$ as functions of the temperature of each layer are different. To simplify the model, uniform thermal parameters are needed. $C_p(T)$ and $\lambda(T)$ of each layer are equivalently calculated, as shown in Equations (6) and (7):

$$C_{eq} = \frac{\sum_N C_n(T) \rho_n t_n}{\sum_N \rho_n t_n} \quad (6)$$

$$\lambda_{eq} = \frac{\sum_N \lambda_n(T) t_n}{\sum_N t_n} \quad (7)$$

where, respectively, C_{eq} and λ_{eq} are equivalent specific heat capacity and thermal conductivity and C_n and λ_n are specific heat capacity and thermal conductivity of each metal layer, and ρ_n is the density and t_n is the thickness of each metal layer. N represents the types of mental materials.

The second part of the heat transfer model is the heat exchange between the outer boundary of the winding and the liquid nitrogen, which is dominated by thermal convection. However, since the increased temperature of the winding causes the liquid nitrogen to boil, the heat transfer model is complicated. In order to simplify the calculation, a heat flux is applied to the outer boundary to simulate the process of boiling liquid nitrogen heat transfer, so that the simulated liquid nitrogen region can be directly deleted. The basic equation is as follows:

$$Q = h(T_s - T_0) \cdot A \cdot (T_s - T_0) \tag{8}$$

where T_s is the temperature of the heat flux boundary, A is the boundary cross-sectional area, T_0 is the liquid nitrogen temperature (77 K), and $h\,(T_s - T_0)$ is the effective nonlinear steady-state convection coefficient (W/(K·m^2)). The value of $h\,(T_s - T_0)$ is liquid nitrogen boiling heat flux curve in the simulation, which is cited in [21].

2.5. Model Convergence and Optimization

2.5.1. Middle Voltage Loop Optimization

As shown in Figure 1, the iterative calculation is divided into three layers: Inner, middle, and outer. The middle voltage loop is used to iterate in the circuit model (calculating the primary and secondary currents) and the electromagnetic model (calculating the resistive voltage drop of the superconducting winding) under a specified supply voltage. The intermediate variable is the superconducting winding resistive voltage drop U_{sc_p}, U_{sc_s}, and the updating method is shown as Equation (9):

$$\begin{cases} U_{sc\_p(k)} = U_{sc\_p(k-1)} - \theta \Delta U_{sc\_p(k)} \\ U_{sc\_s(k)} = U_{sc\_s(k-1)} - \theta \Delta U_{sc\_s(k)} \end{cases} \tag{9}$$

where $U_{sc\_p(k)}$ represents the primary resistive voltage drop at the kth iteration and $U_{sc\_s(k)}$ is the secondary voltage drop, where k refers to the number of iterations in the middle voltage loop; $\Delta U_{sc\_p(k)}$ is the difference between $U_{sc\_p(k)}$ and $U_{sc\_p(k-1)}$; and θ is the learning rate. In order to accelerate convergence in the early stage, a method of exponential decay plus constant is used to calculate θ, such as Equation (10):

$$\theta(k) = \beta e^{-\frac{k}{\tau}} + \xi \tag{10}$$

where β is the exponential decay coefficient, and the value is large (0.2), which ensures that the model can be quickly converged at the beginning; τ controls the decay rate; and ξ is the constant learning rate, which ensures that the learning rate of the model does not decay to a too small value after multiple iterations and cannot update. In the simulation, $\beta = 0.2$, $\tau = 30$, and $\xi = 0.01$.

2.5.2. Inner Voltage Loop Optimization

The inner voltage loop is used to calculate the current distribution in the superconducting and metal layers under a specified total winding current. In the actual calculation, due to the characteristics of the HTS, its current will enter a large nonlinear region when it reaches the critical current, as shown in Equation (1). At this time, if only U_{sc_p} and U_{sc_s} are used as the convergence conditions, the model will oscillate easily and cannot converge.

To solve this problem, when the total winding current is close to or greater than the critical current, U_{metal_p} and U_{metal_s} are used as convergence conditions. The superconducting layer has nonnegligible resistance at this time, a part of current flow through the metal layer, which results in a nonnegligible voltage drop in the metal layer, and this voltage drop can be directly calculated. Because the resistance of the metal layer is constant at the same temperature, there is no oscillation.

Figure 5 shows the number of iterations per step in two cycles (40 ms) before and after optimization. The ordinate represents the iterations of the inner current loop when the outer voltage loop is iterated once; the abscissa is the iterations of the outer voltage loop. Figure 5 shows that the number of iterations at some special points can be greatly reduced after optimization.

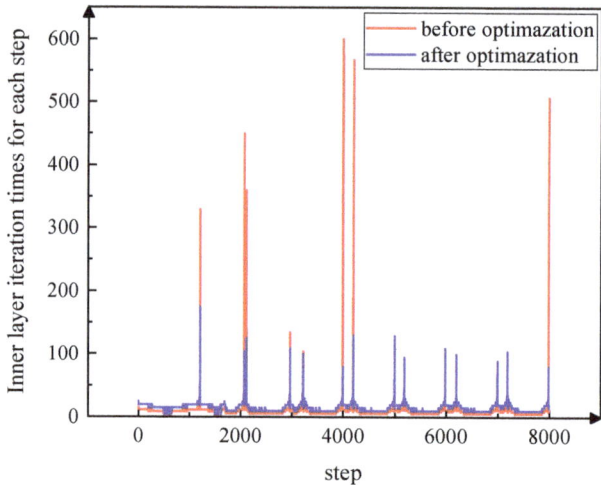

Figure 5. Number of iterations at each step before optimization and after optimization.

3. Results and Discussion

3.1. Typical Examples

3.1.1. Current and Voltage

The primary and secondary current curves at different supply voltages were obtained and used to analyze the macroscopic operating characteristics of the SFCL in the simulation. The operating conditions under two typical voltages (voltage amplitudes of 0.5 V, 30 V) are listed, including primary current, primary superconducting layer current, secondary current, secondary superconducting layer current, and voltage.

Figure 6 shows the simulation results of the voltage and current curves at a voltage magnitude of 0.5 V. The current flows from the superconducting layer before the primary current reaches 60 A for the first time; after the current exceeds 60 A, the shunting phenomenon begins to appear. It shows that the superconducting layer begins to show obvious resistance because its current exceeds the critical current. However, compared with the metal layer resistance, this resistance is still small, so only a few amperes of current flow through the metal layer.

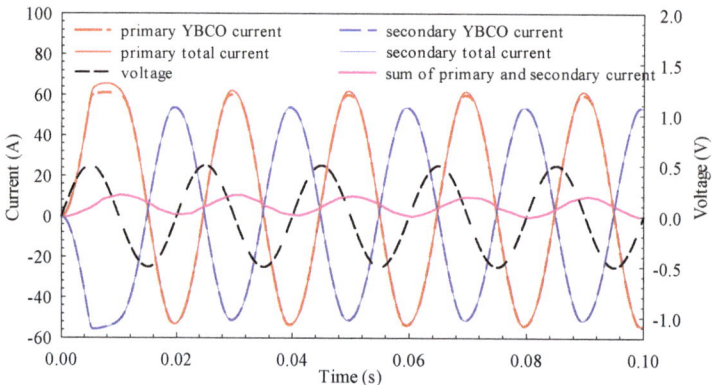

Figure 6. Simulation results of voltage and current curves at voltage magnitude of 0.5 V.

When the waveform is stable, the primary current amplitude is 53.9 A, the secondary current amplitude is 51.2 A, and the phase difference between the two is 174.6°. In order to explain the shielding effect of the secondary winding on the magnetic flux generated by the primary winding, the sum of the primary and secondary currents is defined as "net current," which generates magnetic flux (the ratio of primary to secondary is 1:1). The purple curve in Figure 6 is net current; the amplitude is 5.1 A and bias is 6.0 A, which is 9.5% of the primary current. The secondary winding shields most of the magnetic flux generated by the primary winding. Only 9.5% of the primary current builds an alternating magnetic field, so the overall leakage inductance is close to zero. The primary and secondary currents flow through the HTS layer, so the overall impedance of the SFCL is close to zero. The phase difference between the primary current and the voltage waveform is 84.6°. The leakage inductance of the primary and secondary windings makes the SFCL inductive.

Figure 7 shows the simulation results of the voltage and current curves at a voltage magnitude of 30 V. The first peak of the primary current is 652.7 A, and the current wave becomes steady state in the fifth cycle with the peak value of 427.4 A. After the current of the secondary winding becomes stable, the peak value is 127.1 A, and the amplitude of net current generating the alternating magnetic field is 399.5 A, which is 93.5% of the primary current. It indicates that most of the primary current is used to construct the alternating magnetic field, and the secondary current can only shield a fraction of the magnetic flux generated by the primary current. The SFCL has large inductance; meanwhile, the resistive component of the winding is also increased due to the large primary current. The phase difference between the primary current and the voltage is 57.6° in steady state.

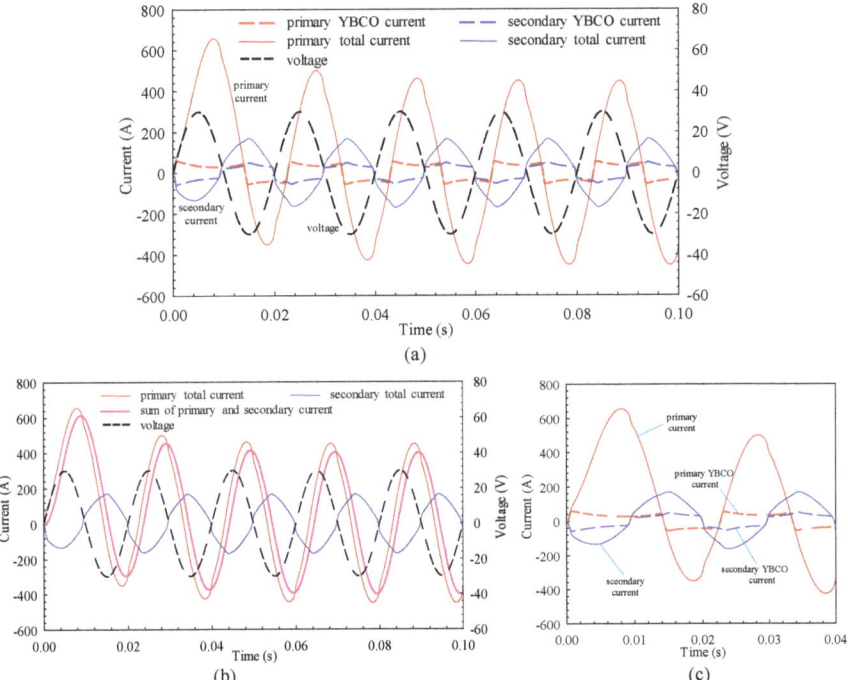

Figure 7. Simulation results of voltage and current curves at voltage magnitude of 30 V: (**a**) Overall curves, (**b**) net current that generates magnetic flux, (**c**) details of curves at first two cycles.

At this moment, due to the large primary and secondary current, a considerable amount of joule heat is generated, and the influence of the increased winding temperature on the HTS strip cannot be ignored. As shown in Figure 7a, the heat accumulation results in decreased saturation currents in the primary and secondary superconducting layers, because the critical current decreases as the winding temperature rises, and the equivalent impedance increases. The temperature rise curve and temperature distribution of the winding will be analyzed in detail later.

3.1.2. Magnetic Field and Electrodynamic Force

By analyzing the distribution of the magnetic field, the distribution of critical current can be derived, then the distribution of power density can be inferred. In addition, the maximum electrodynamic force acting on the limiter depends on the first peak of primary current under short-circuit conditions. Figure 8 shows the magnetic field distribution at the first current peak of the primary winding under a voltage magnitude of 30 V.

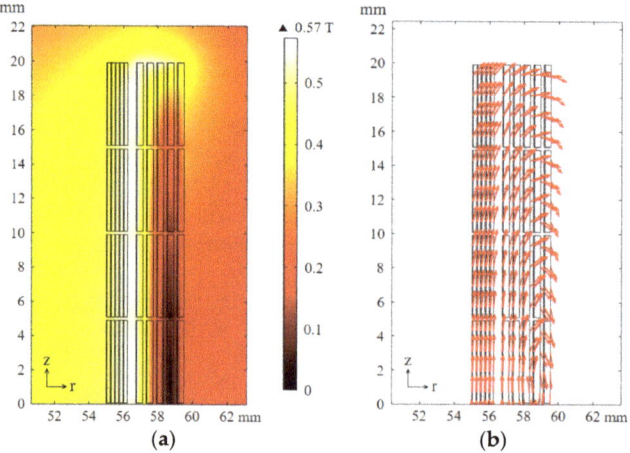

Figure 8. Magnetic field distribution at first current peak of primary winding under voltage magnitude of 30 V: (**a**) Flux density norm value distribution, (**b**) magnetic field direction diagram.

In the area between primary and secondary windings, the entire secondary winding, and the inside and ends of the primary winding, the magnetic field is largest. Among them, the secondary winding and inside and ends of the primary winding are mainly influenced by the magnetic field parallel to the tape. According to Equation (2), this magnetic field has little effect on the critical current, which results in a low power loss density; the ends of the primary winding are mainly affected by the magnetic field vertical to the tape, and this results in low critical current density and high power loss density. To improve the recovery speed after cutting off the short fault, the above problem should be taken into consideration in the design of a cooling structure. If necessary, tapes with higher critical current and better cooling structure can be used in the ends of the primary winding.

Figure 9 shows the electrodynamic force distribution at the first current peak of the primary winding under a voltage magnitude of 30 V. The electrodynamic force affecting the secondary winding is small and the direction is negative; in the primary winding, the inner side is influenced by positive force and the outer side, and the winding is squeezed in the r-axis direction. For electrodynamic force in the z-axis direction, the secondary winding has a small influence and the direction is positive; in the primary winding, the end is greatly affected by a negative force and the middle is influenced by a positive force, and the winding is squeezed in the z-axis direction.

As shown in Figure 9c, different parts of the winding are influenced by electrodynamic force in different directions. Electrodynamic force is a macro performance of the Lorentz force; the magnetic field of the conductor is generated by its own current, so the direction of electrodynamic force does not change with the direction of the current.

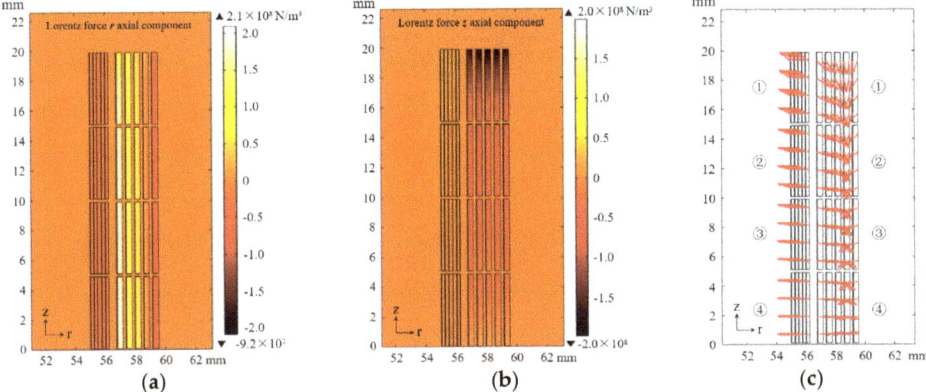

Figure 9. Electrodynamic force distribution at the first current peak of primary winding under voltage magnitude of 30 V: (**a**) R-axial component, (**b**) z-axial component, (**c**) electrodynamic force direction.

To ensure that the SFCL will not be destroyed in the structure under short-circuit current, the electrodynamic force affecting different parts of the winding should be considered in the design. The superconducting coils in the primary and secondary windings of the upper half of the SFCL are numbered and calculated by the electrodynamic force they affect; the numbers are shown in Figure 9c and the calculation results are shown in Table 2.

Table 2. Stress of pancake winding under first peak current of primary winding.

	Primary Winding	Secondary Winding
No.	Electrodynamic force r-axial component (N/m)	
1	438.3	−116.6
2	493.4	−138.6
3	517.7	−144.9
4	528.4	−147.1
No.	Electrodynamic force z-axial component (N/m)	
1	−1023.4	75.9
2	−535.8	45.8
3	−283.8	24.7
4	−89.9	7.9

The force acting on the secondary winding is not large, and the external shedding force in the z-axis should be considered in the design; the force acting on the primary winding is large, more attention should be paid to the outward force, and it is necessary to enhance the structure because of the large force.

3.1.3. Temperature

When the short-circuit current is large, a large amount of joule heat generated by winding will result in a significant temperature rise, affecting the electromagnetic properties of the tape. Taking the simulation result voltage magnitude of 30 V as an example, the rising temperature curve of primary and secondary windings is shown in Figure 10. Since the rising temperatures of different turns are different in the same winding, the averages of temperature rise of the primary and secondary sides are used. As shown in Figure 10, the average temperature of the primary winding increases from the initial 77 K to 85.8 K, and that of the secondary winding increases from 77 K to 81.1 K. Figure 6 shows that the primary current is greater than the secondary current, so the primary temperature rise should be greater than the secondary, but because the primary side resistivity is less than the secondary side, the combined action causes a minor difference in temperature rise.

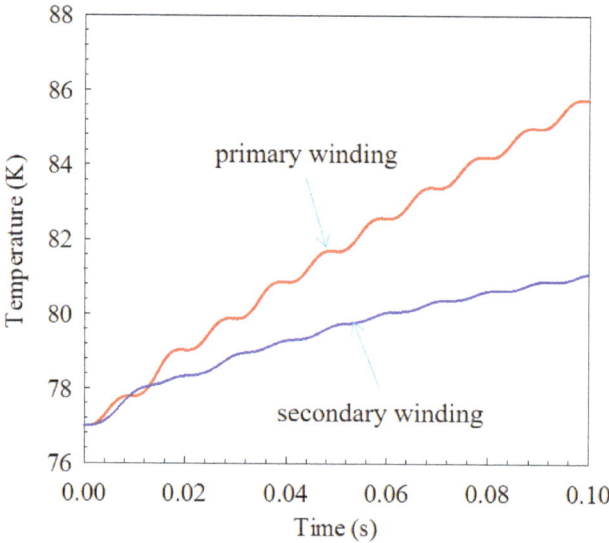

Figure 10. Rising temperature of primary and secondary windings under voltage magnitude of 30 V.

In order to further analyze the internal operation of the SFCL, Figure 11 shows the temperature distribution of the primary and secondary windings at 0.02 s, 0.04 s, 0.06 s, and 0.08 s (only the upper half of the winding is drawn because it is symmetric).

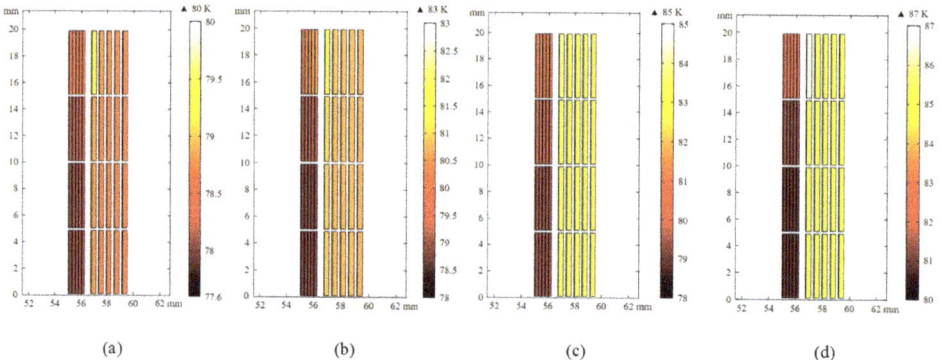

Figure 11. Temperature distribution of primary and secondary windings at (**a**) 0.02 s, (**b**) 0.04 s, (**c**) 0.06 s, and (**d**) 0.08 s.

As shown in Figure 11, the portion with the greatest temperature rise is at the end of the primary innermost winding. Since it is affected by the largest magnetic field (especially the magnetic field perpendicular to the tape), the critical current there is minimized. A large amount of joule heat causes the greatest temperature rise. In the space between the primary and secondary windings, due to the opposite direction of the current, the generated magnetic fields are superimposed and enlarged, but the magnetic field direction is parallel to the strip, so the field has little effect on the critical current. It can be seen that the temperature rises more inside the primary winding than outside. Similarly, in the secondary winding, the temperature rises the most at the end. Since the structure of all-metal and no-insulation layers makes it possible to exchange heat with liquid nitrogen directly, the heat conduction speed is fast. So overall, the temperature rises significantly less in the secondary winding than in the primary winding.

As shown in Figure 11, compared with the secondary winding, the temperature of the primary winding rises more greatly in the current-limiting state of the HTS FCL. In order to speed up the recovery time after the fault current is cut off, the heat dissipation problem should be considered in the actual design and manufacture of the primary winding structure. While ensuring support strength, the contact area between the winding and the liquid nitrogen is increased as much as possible.

3.2. Short-Circuit Test

In order to simulate the fault conditions in an actual grid, a SFCL prototype with the same simulation structure is fabricated and connected to the short-circuit test platform. The primary winding is an insulation superconducting winding with insulating material of KAPTON, and the secondary winding is a no-insulation superconducting winding. In this paper, superconducting winding is wound with second-generation high temperature superconducting material—YBCO tapes with stainless-steel package, which are produced by Shanghai Superconductor.

HTS coated conductor (CC) tapes were used in this paper is 4.8 mm wide and critical current is 40 A (77 K, self-field). Superconducting tapes are with a 1.5 μm thick YBCO layer, a 50 μm thick Hastelloy substrate, a 2 μm thick silver cap layer, and a 10 μm + 10 μm thick copper stabilization layer. The structure diagram of YBCO tape is shown in Figure 12. Then, the outermost is packaged in 75 m thick stainless steel.

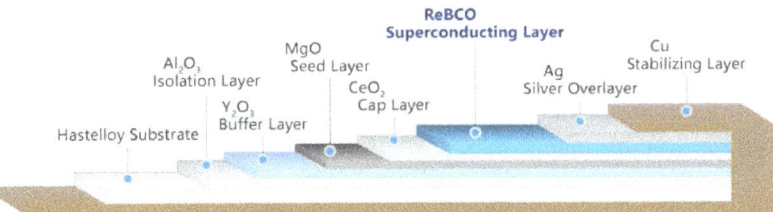

Figure 12. Structure diagram of YBCO tape.

The primary winding and secondary winding consist of four double-pancake coils, respectively. A pancake coil is a coil with a flat spiral form, and a double-pancake coil is two pancake coils made by a superconducting tape. Double-pancake coils are connected in series by soldering, and resistance is 1.5 μΩ. Specifications of the SFCL prototype is shown in Table 1, same with the SFCL model, and the inductive SFCL prototype is shown in Figure 13.

Critical current of primary winding and secondary winding is 28 A and 30 A (77 K), respectively. Self-inductance of primary winding and secondary winding is 0.237 mH and 0.206 mH. The mutual inductance between the primary and the secondary is 0.196 mH, and the coupling coefficient is 0.887.

The short-circuit test platform is shown in Figure 14; voltage source consists of voltage regulator (input: 400 V, output: 0~400 V) and step-down transformer (ratio is 20:1 or 10:1). Different voltage can be obtained by changing output voltage of voltage regulator. A load resistor connected in parallel with IGBT fast switching is in series with the SFCL prototype, and the IGBT controlled by the host computer is used to generate a short-circuit voltage for a specified duration.

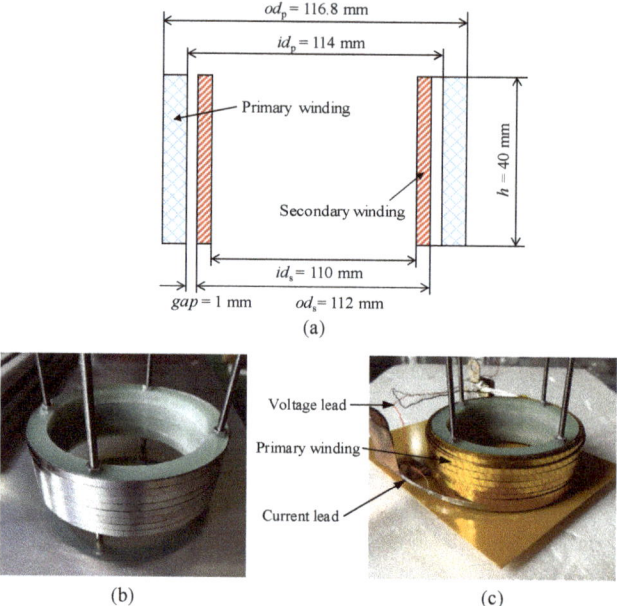

Figure 13. Superconducting inductive SFCL prototype: (**a**) Schematic diagram, (**b**) secondary winding, (**c**) SFCL prototype.

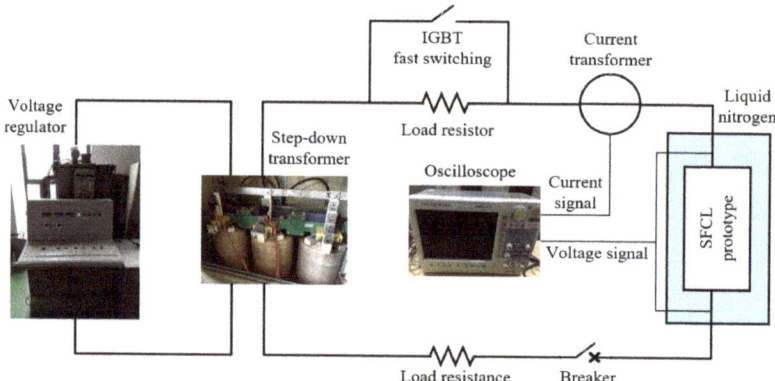

Figure 14. Short-circuit test platform.

The step-down transformer is used as a voltage source and the turn-off time is controlled to provide a short-circuit current with a fixed time for the SFCL. In this experiment, the short-circuit time is 5 cycles of 100 ms.

Considering the difficulty of measuring and simulating the internal impedance of voltage regulator, the experimental voltage value is directly input into the transient model for calculation. Figure 15 shows the comparison of the experimental waveform (left) and the simulated waveform (right) at different voltages (waveforms in two cycles after the short-circuit current becomes steady state).

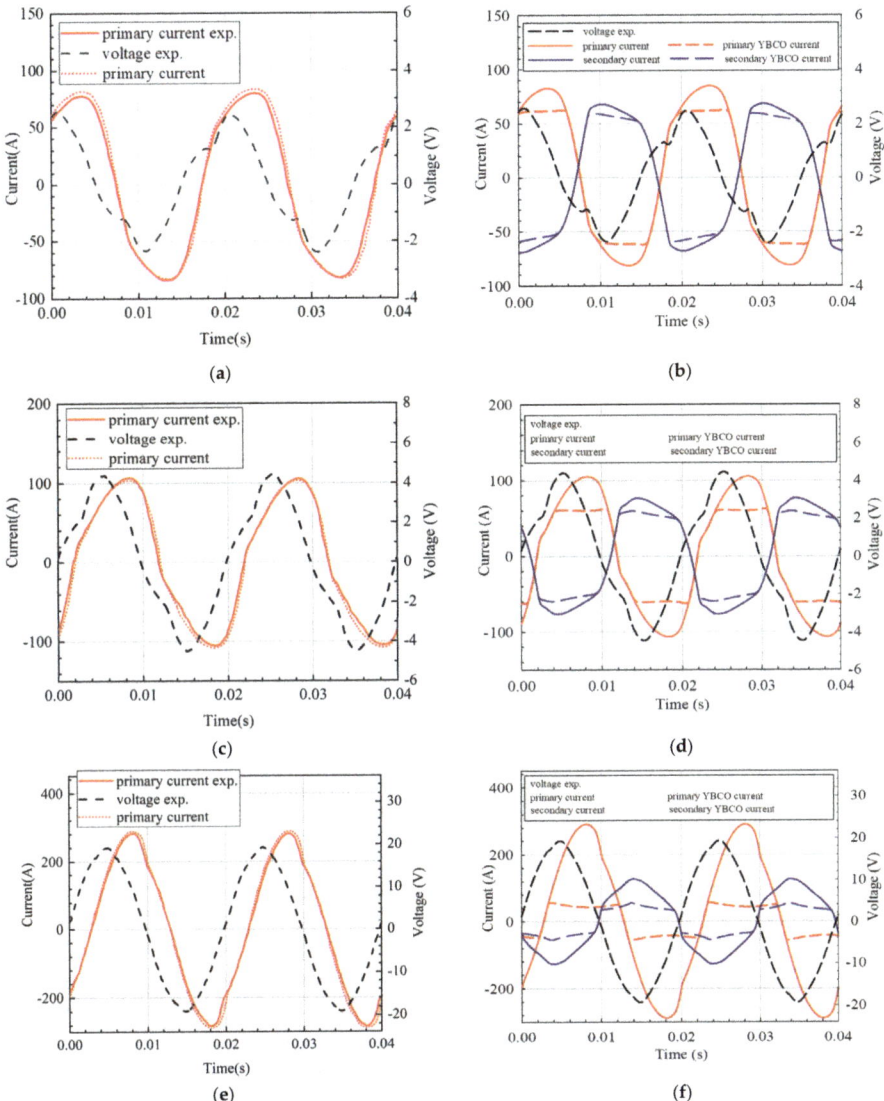

Figure 15. Comparison of experimental (left) and simulation (right) results of voltage and current curves. (**a**) Comparison of primary current curve between experimental and simulation under voltage magnitude of 2.6 V, (**b**) overall simulation current under voltage magnitude of 2.6 V, (**c**) comparison of primary current curve between experimental and simulation under voltage magnitude of 4.5 V, (**d**) overall simulation current under voltage magnitude of 4.5 V, (**e**) comparison of primary current curve between experimental and simulation under voltage magnitude of 19.5 V, and (**f**) overall simulation current under voltage magnitude of 19.5 V.

As shown in Figure 15, the simulated current waveform is basically consistent with the experimental waveform, which proves that the simulation can well simulate the operating characteristics of the air core current limiter under short-circuit current, which lays a foundation for future SFCL design.

The error between the experiment and the simulation may be caused by the measurement error of the strip electrical and thermal parameters, or by nonuniform electrical and thermal parameters of the strip in the longitudinal direction. The primary current waveform is related to the secondary current waveform, especially when the secondary current begins to shunt in the HTS and metal layers, the secondary side impedance increases, and the instantaneous current value begins to decrease compared with the standard sine wave. At the same time, the primary current is distorted, and the overall distortion is less than the secondary winding.

4. Conclusions

In this paper, a numerical study of transient states for the process of limiting current in inductive FCLs is proposed. The model is based on the coupling of multiphysics finite element simulation and a circuit model. By optimization, the number of iterations is reduced, and operational speed is greatly increased. The overall model calculation process, submodel setting, and analysis of the macroscopic characteristics (voltage, current) and microscopic characteristics (distribution of magnetic field, electrodynamic force, and temperature) of the SFCL under the current-limiting state are introduced. In SFCL design, more attention should be paid to the large outward force acting on the primary winding, and it is necessary to enhance the structure; the heat dissipation problem should be considered, while ensuring that the support strength and the contact area between the winding and the liquid nitrogen are increased as much as possible. The short-circuit experimental data of an air core inductive superconducting current limiter (SFCL) prototype was compared with the simulation results to verify the reliability of the simulation and provides a theoretical basis for the future design and manufacture of SFCL. Further, inductive SFCL applied in a short test of a large capability transformer under the grid operating condition can be designed. In this paper, some parameters in electromagnetic model and heat transfer model can be further refined and the simulation accuracy is improved. Similarly, better optimization method can be used to accelerate convergence.

Author Contributions: Conceptualization, J.S.; methodology, J.S. and W.L.; validation, D.Q.; data curation, J.C., H.Y.; writing—original draft, W.L.; writing—review and editing, W.L., D.Q., J.C., H.Y.; supervision, Z.H.

Funding: This research was funded by the National Natural Science Foundation of China, grant number 51807118.

Conflicts of Interest: The authors declare no conflict of interest.

References

1. Ruiz, H.S.; Zhang, X.; Coombs, T.A. Resistive-type superconducting fault current limiters: Concepts, materials, and numerical modeling. *IEEE Trans. Appl. Supercond.* **2015**, *25*, 5601405. [CrossRef]
2. Naeckel, O.; Noe, M. Design and Test of an Air Coil Superconducting Fault Current Limiter Demonstrator. *IEEE Trans. Appl. Supercond.* **2014**, *24*, 5601605. [CrossRef]
3. Meerovich, V.; Sokolovsky, V.; Bock, J.; Gauss, S.; Goren, S.; Jung, G. Performance of an inductive fault current limiter employing BSCCO superconducting cylinders. *IEEE Trans. Appl. Supercond.* **1999**, *9*, 4666–4676. [CrossRef]
4. Usoskin, A.; Mumford, F.; Dietrich, R.; Handaze, A.; Prause, B.; Rutt, A.; Schlenga, K. Inductive Fault Current Limiters: Kinetics of Quenching and Recovery. *IEEE Trans. Appl. Supercond.* **2009**, *19*, 1859–1862. [CrossRef]
5. Kozak, J.; Majka, M.; Kozak, S.; Janowski, T. Design and tests of coreless inductive superconducting fault current limiter. *IEEE Trans. Appl. Supercond.* **2012**, *22*, 5601804. [CrossRef]
6. Paul, W.; Lakner, M.; Rhyner, J.; Unternährer, P.; Baumann, T.; Chen, M.; Guerig, A. Test of 1.2 MVA high-superconducting fault current limiter. *Supercond. Sci. Technol.* **1997**, *10*, 914–918. [CrossRef]
7. Kozak, J.; Majka, M.; Kozak, S. Experimental Results of a 15 kV, 140 A Superconducting Fault Current Limiter. *IEEE Trans. Appl. Supercond.* **2017**, *27*, 5600504. [CrossRef]
8. Cvoric, D.; De Haan, S.W.H.; Ferreira, J.A.; Van Riet, M.; Bozelie, J. Design and Testing of Full-Scale 10 kV Prototype of Inductive Fault Current Limiter with a Common Core and Trifilar Windings. In Proceedings of the 2010 International Conference on Electrical Machines and Systems, Incheon, South Korea, 10–13 October 2010.

9. Hong, Z.; Campbell, A.M.; Coombs, T.A. Numerical solution of critical state in superconductivity by finite element software. *Supercond. Sci. Technol.* **2006**, *19*, 1246–1252. [CrossRef]
10. Hong, Z.; Coombs, T.A. Numerical Modelling of AC Loss in Coated Conductors by Finite Element Software Using H Formulation. *J. Supercond. Nov. Magn.* **2010**, *23*, 1551–1562. [CrossRef]
11. Zhang, H.; Zhang, M.; Yuan, W. An efficient 3D finite element method model based on the T–A formulation for superconducting coated conductors. *Supercond. Sci. Technol.* **2017**, *30*, 024005. [CrossRef]
12. Wang, Y.; Zhang, M.; Grilli, F.; Zhu, Z.; Yuan, W. Study of the magnetization loss of CORC cables using 3D TA formulation. *Supercond. Sci. Technol.* **2019**, *32*, 025003. [CrossRef]
13. Kozak, S.; Janowski, T.; Wojtasiewicz, G.; Kozak, J.; Kondratowicz-Kucewicz, B.; Majka, M. The 15 kV Class Inductive SFCL. *IEEE Trans. Appl. Supercond.* **2010**, *20*, 1203–1206. [CrossRef]
14. Naeckel, O.; Noe, M. Conceptual Design Study of an Air Coil Fault Current Limiter. *IEEE Trans. Appl. Supercond.* **2013**, *23*, 5602404. [CrossRef]
15. Hekmati, A.; Vakilian, M.; Fardmanesh, M. Flux-Based Modeling of Inductive Shield-Type High-Temperature Superconducting Fault Current Limiter for Power Networks. *IEEE Trans. Appl. Supercond.* **2011**, *21*, 3458–3464. [CrossRef]
16. Kado, H.; Ichikawa, M.; Ueda, H.; Ishiyama, A. Trial Design of 6.6kV Magnetic Shielding Type of Superconducting Fault Current Limiter and Performance Analysis Based on Finite Element Method. *Electr. Eng. Jpn.* **2010**, *145*, 50–57. [CrossRef]
17. Sheng, J.; Wen, J.; Wei, Y.; Zeng, W.; Jin, Z.; Hong, Z. Voltage-Source Finite-Element Model of High Temperature Superconducting Tapes. *IEEE Trans. Appl. Supercond.* **2015**, *25*, 7200305. [CrossRef]
18. Sheng, J.; Hu, D.; Ryu, K.; Yang, H.S.; Li, Z.Y.; Hong, Z. Numerical Study on Overcurrent Process of High-Temperature Superconducting Coated Conductors. *J. Supercond. Nov. Magn.* **2017**, *30*, 3263–3270. [CrossRef]
19. Grilli, F.; Sirois, F.; Zermeno, V.M.R.; Vojenciak, M. Selfconsistent modeling of the Ic of HTS devices: How accurate do models really need to be? *IEEE Trans. Appl. Supercond.* **2014**, *24*, 8000508. [CrossRef]
20. Wang, Y.; Chan, W.K.; Schwartz, J. Self-protection mechanisms in no-insulation (RE)Ba$_2$Cu$_3$O$_x$ high temperature superconductor pancake coils. *Supercond. Sci. Technol.* **2016**, *29*, 045007. [CrossRef]
21. Sakurai, A.; Shiotsu, M.; Hata, K. Boiling heat transfer characteristics for heat inputs with various increasing rates in liquid nitrogen. *Cryogenics* **1992**, *32*, 421–429. [CrossRef]

© 2019 by the authors. Licensee MDPI, Basel, Switzerland. This article is an open access article distributed under the terms and conditions of the Creative Commons Attribution (CC BY) license (http://creativecommons.org/licenses/by/4.0/).

Article

Modeling of High-T_c Superconducting Bulk using Different J_c–T Relationships over Dynamic Permanent Magnet Guideway

Ye Hong, Jun Zheng * and Hengpei Liao

Applied Superconductivity Laboratory, State Key Laboratory of Traction Power, Southwest Jiaotong University, Chengdu 610031, China
* Correspondence: jzheng@swjtu.edu.cn

Received: 31 July 2019; Accepted: 6 September 2019; Published: 9 September 2019

Abstract: The linear temperature dependence of critical current density $J_c \propto ((T_c-T)/(T_c-T_0))$ and the nonlinear functions of $J_c \propto (1-(T/T_c)^2)^\alpha$ with the exponent α equal to 1, 3/2, and 2 are used to calculate the dynamic levitation force, the temperature distribution, and the current density distribution of the high-temperature superconducting (HTS) YBaCuO bulk over a permanent magnetic guideway (PMG). The calculations were based on the H-formulation and E–J power law. The model of the HTS bulk and the PMG has been built as a geometric entity by finite element software. To simulate the magnetic field fluctuation caused by the PMG arrangement irregularity, a small amplitude vibration in the vertical direction is applied to the PMG during the calculations. Both the low vibration frequency of 2 Hz and the high vibration frequency of 60 Hz are analyzed as the representative converted linear speeds of 34 km/h and 1018 km/h for magnetic levitation (Maglev) application. We compared the electromagnetic-thermo-force modeling with the experiments and the previous model without considering the thermal effect. The levitation force computed by the J_c–T relationship, in which J_c is proportional to $(1-(T/T_c)^2)^2$, is found to be in best agreement with the experimental data under quasi-static conditions. This work can provide a reference for the HTS electromagnetic-thermal-force coupling reproduction method of HTS Maglev at high speed.

Keywords: high-temperature superconducting bulk; modeling; magnetic levitation; electromagnetic-thermo-force coupling; high speed

1. Introduction

Due to the inherent flux-pinning effect, high-temperature superconducting (HTS) bulk has been potentially used in HTS magnetic levitation (Maglev) transportation [1–4] and many other applications [5,6]. To investigate the electromagnetic characteristic of the HTS Maglev system, a proper HTS E-J constitutive law is necessary. For the HTS materials, the critical current density changes along the temperature. In the past, Matsushita et al. [7] investigated the single-grain YBaCuO specimen by measuring the critical current density J_c–T, and proposed the current density as a function of $1-(T/T_c)^2$. Yamasaki et al. [8] reported that J_c is proportional to $(1-AT+BT^2)$, where A and B were constants for the HTS material of Bi-2223 thin films. Another research study [9] described J_c as a function of $(1-T/T_c)^\gamma$ based on the exploration of Bi-2212/Ag wires. In addition, $J_c \propto (1-(T/T_c)^2)^\alpha$ for Bi-2212 tapes was mentioned in the book [10]. For 2G HTS YBCO bulks, Braeck [11] assumed a linear temperature dependence of the critical current as $J_c \propto ((T_c-T)/(T_c-T_0))$; and then Tsuchimoto [12] presented an exact nonlinear relationship of $J_c \propto (1-(T/T_c)^2)^2$.

Later, from the point of view of superconducting YBaCuO application, Tsukamoto et al. [13] employed the linear temperature dependence of the critical current density to calculate the temperature variation and the trapped magnetic field in YBCO bulks under an AC external field. Using this linear formula, Tixador et al. [14] calculated the current distribution and AC losses of the YBCO slab. In recent years, Ye [15] and Huang [16] studied the dynamic thermal effect of the HTS Maglev system using YBaCuO bulks by the linear J_c–T relation.

In this paper, in order to elucidate which J_c–T relationship is more appropriate to model the HTS Maglev system, we used four different J_c–T functions to calculate the dynamic levitation force, the temperature distribution, and the current density distribution of the HTS bulk over the applied permanent magnetic guideway (PMG). The modeling characterized by the H-formulation and the E–J power law was implemented in the finite element software COMSOL Multiphysics 5.3a. With the magnetic field formulation (MFH) interface in the AC/DC module, the calculation subdomains of the HTS bulk and the PMG were built as the geometric entity. During the calculations, a vertical vibration with small amplitude was applied to the PMG to simulate the magnetic field fluctuation caused by the inevitable PMG irregularity. Different from the previous modeling [17,18], the thermal effect was taken into account by coupling the heat transfer module. To study the levitation performance of the HTS Maglev system at high speed, a vibration with the frequency of 60 Hz was set to simulate the magnetic field inhomogeneity, which is equivalent to the linear velocity of about 1018 km/h of the circular PMG employed in the experiments. The converted linear velocity v of the circumferential PMG in the experiments can be expressed as:

$$v\,(\text{km/h}) = r(\text{m}) \times 2\pi n(\text{rpm}) \times \frac{60}{1000} \quad (1)$$

where r is the radius of the circular PMG, which is 0.75 m; the rotation rate of the PMG, n (rpm), is derived from the vibration frequency of the guideway set in the simulation. This occurs since one cycle of the magnetic field fluctuation during the dynamic modeling is approximately tantamount to the real PMG magnetic field, which the HTS bulk subjects to when the circular guideway rotates for one circle in the experiment [19,20]. Thus, the cycle of the magnetic field fluctuation is 1/60 s when the frequency is 60 Hz, and the PMG's rotation rate n is 3600 rpm. In that case, the calculation results by different J_c–T functions showed the levitation force attenuation during the vibration process. In comparison with the measurements in this paper, the J_c–T relationship in which J_c is proportional to $(1-(T/T_c)^2)^2$ shows better agreement with the experiment.

2. Theoretical Model

The HTS bulk and the opposite -polar-arranged PMG were built as the geometric entity in the COMSOL. The remanence of the permanent magnet B_r was set at 0.8 T. With the MFH (magnetic field formulation) interface, we can easily simulate the magnetic field produced by any complex- shaped magnets. In this study, after the PMG reached the working height (WH), a small amplitude vertical vibration was applied to the PMG to simulate the magnetic field fluctuation. The first movement stage is regarded as the quasi-static process, since the speed is as low as 1 mm/s, while the following vibration process is the dynamic levitation stage.

Figure 1 shows the computation subdomains of the dynamic model and the movement process of the PMG. In the modeling, the PMG first moves from the field-cooling height (FCH = 30 mm) to the working height (WH = 15 mm) at 1 mm/s. This process takes 15 s. Afterwards, it relaxed from 15 to 120 s (Relaxation I) to let the flux fully redistribute. Then, a sine function with 1-mm amplitude is applied to the PMG for 20 s. After the dynamic condition, the second relaxation process (Relaxation II) takes place, which is from 140 s to the end. In the dynamic condition, the vibration frequency conditions of 2 and 60 Hz are calculated, respectively. According to the size of the circular PMG in the experimental equipment SCML-03 [21], the vibration frequencies of 2 and 60 Hz are corresponding to the linear velocity of 34 and 1018 km/h for the Maglev above the PMG, respectively.

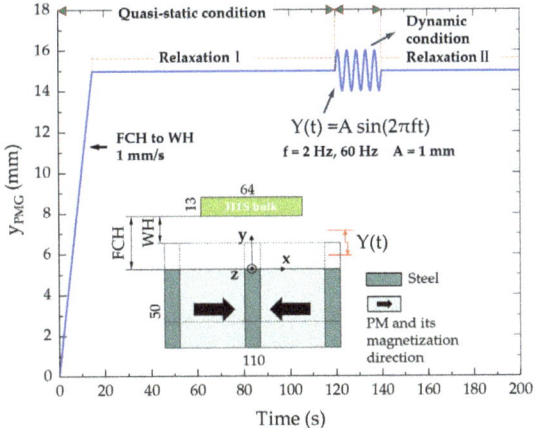

Figure 1. Vertical displacement variation of the permanent magnetic guideway (PMG) and two-dimensional (2D) model for the high-temperature superconducting (HTS) Maglev system. Vibration amplitude A = 1 mm. Field-cooling height (FCH) = 30 mm, working height (WH) = 15 mm. The quasi-static condition from 0 s to 120 s includes the 0–15 s upward movement from FCH to WH and the Relaxation I process from 15 to 120 s. Then, after 20 s of vibration, the second relaxation process (Relaxation II) happened from 140 s to the end.

Equations

To analyze the electromagnetic characteristics of the HTS bulk, the *H*-formulation and the well-known Maxwell equations (neglecting the displacement current term) [22] are as follows:

$$\mu \frac{\partial H}{\partial t} + \nabla \times E = 0 \tag{2}$$

$$E = \rho J \tag{3}$$

$$J = \nabla \times H \tag{4}$$

where *H* is the magnetic field strength; ρ is the resistivity of the material, which is assumed to be isotropic in the two-dimensional (2D) model; *E* is the electric field; and *J* is the current density. Equations (2)–(4) are solved by the MFH interface in the AC/DC module.

The strong nonlinear relationship between *E* and *J* of the HTS material can be characterized by the experimental empirical *E–J* power law [23]:

$$E = E_c \left(\frac{J}{J_c}\right)^m \tag{5}$$

where E_c is the critical current criterion equal to 100 μV/m, and *m* is the power law exponent, which is usually 21. The parameter J_c in the constitutive law depends on the temperature *T*. Thus, the thermal effect is coupled in the modeling using the heat transfer module. Four J_c–*T* relationships are considered.

(1). The linear dependence [15]:

$$J_c = J_{c0} \frac{T_c - T}{T_c - T_0} \tag{6}$$

where J_{c0} is the critical current density of the HTS bulk at $T = T_0$ (77 K); T_c is the critical temperature; and T_0 is the coolant temperature.

(2). The nonlinear dependence with $\alpha = 1, 3/2, 2$ turns into [10]:

$$J_c = J_{c1}\left(1 - \left(\frac{T}{T_c}\right)^2\right)^{\alpha} \qquad (7)$$

where the critical current density J_{c1} is obtained by Equation (6) with $T = 0$ K. In addition, we calculated the case without considering the thermal effects. In that case, J_c as a constant equals J_{c0}.

The thermal equilibrium equation is expressed as:

$$C_p \frac{\partial T}{\partial t} - \nabla \cdot (\lambda \nabla T) = EJ \qquad (8)$$

where C_p is the heat capacity per unit volume of the superconductor; and λ is the thermal conductivity of the superconductor.

The levitation force is obtained by the Lorenz force formula at each time instant as:

$$F_y(t) = \int_S B \times J \, dS \; \left[\frac{N}{m}\right] \qquad (9)$$

where S is the cross-section of the HTS bulk in the x-y plane. We assume four bulks placed along the z-direction to get the larger force; the total levitation force equals F_y (in the actual calculation, F_y is the force density along the length at the z-direction) times 128 mm, since each HTS bulk is 32-mm wide along the z-direction and there are four bulks in the experiments. Table 1 summarizes the simulation parameters. The material properties of the HTS bulk are based on the melt textured three-seeded rectangular YBaCuO bulk made by the ATZ GmbH (Torgau, Germany).

Table 1. Parameters for the modeling.

Symbol	Value	Name
E_c	1×10^{-4} V/m	Critical current criterion
J_{c0}	1.1×10^8 A/m^2	Critical current density
m	21	Power law exponent
B_r	0.8 T	Remanence of the PM
T_c	92 K	Critical temperature
T_0	77 K	Initial temperature
C_p	132 J/(kg·K)	Heat capacity per unit volume
λ	4 W/(m·K)	Thermal conductivity

3. Results

Figure 2 displays the normalized quasi-static levitation force obtained by calculation and experiment. In the experiment, four rectangular three-seeded YBaCuO bulks, fabricated by ATZ GmbH in Germany, were mounted in a sample holder fixed above the circular PMG of the SCML-03. The bulk size is $64 \times 32 \times 13$ mm^3. The cross-section of the PMG is the same as the geometry in the simulation, which is shown in Figure 1. The experiment process was the same as the quasi-static process of Figure 1. The YBaCuO bulks were first field cooled with a height of 30 mm, which is the distance between the top surface of the PMG and the bottom of the YBaCuO bulks. Afterwards, the bulks were brought down at 1 mm/s from the FCH to the WH (15 mm), and then relaxed for 10 min.

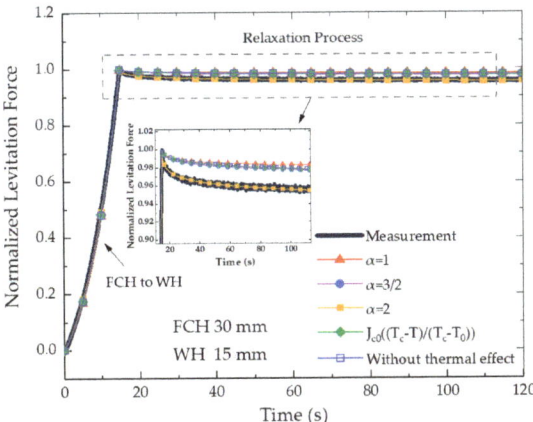

Figure 2. Normalized quasi-static levitation force versus time under different J_c–T relationships of: $J_c = J_{c1}(1-(T/T_c)^2)^\alpha$, $\alpha = 1, 3/2, 2$; $J_c = J_{c0}((T_c-T)/(T_c-T_0))$; $J_c = J_{c0}$. The PMG in the simulation (or bulks in the experiment) moved from the FCH (30 mm) to the WH (15 mm) at the vertical 1 mm/s speed, and then relaxed for over 100 s. The original values of the levitation force at 15 s were 126.40 N (measured), 224.40 N ($\alpha = 1$), 216.58 N ($\alpha = 3/2$), 203.62 N ($\alpha = 2$), 216.52 N ($J_{c0}((T_c-T)/(T_c-T_0))$), and 216.51 N ($J_c = J_{c0}$), respectively.

The numerical results are computed based on the J_c–T functions of Equations (6) and (7) with $\alpha = 1, 3/2$, and 2, as well as the case without considering thermal effect. The levitation forces obtained by calculation and measurement are normalized by dividing its maximum force at 15 s. The original values of the levitation force at 15 s were 126.40 N (measured), 224.40 N ($\alpha = 1$), 216.58 N ($\alpha = 3/2$), 203.62 N ($\alpha = 2$), 216.52 N ($J_{c0}((T_c-T)/(T_c-T_0))$), and 216.51 N ($J_c = J_{c0}$), respectively. So, all the normalized levitation forces equaled 1 only at 15 s, to better compare the difference between the force trends.

It is noted that the levitation force increases gradually during the PMG moving from the field-cooling height to the working height. Then, it shows clear attenuation at the first few seconds during the relaxation process. This is due to the great change of the external magnetic field caused by the large-range movement of the PMG or the HTS bulk, which leads to the redistribution of the flux inside the bulk. The inset in Figure 2 is the partial enlargement of the maximum force region. From the inset, we can see that the calculated levitation force with α equals 2 by Equation (7), which agrees best with the measurements in the quasi-static condition.

Figure 3 further displays the normalized dynamic levitation forces from 115 to 155 s at 2 and 60 Hz by five different calculations of different J_c–T formulas. Figure 3a,c show the general view of the dynamic process, while Figure 3b,d zoom in the vibration end to better compare the different levitation force changes. The corresponding movement of the PMG is depicted in Figure 1. It is found that the continuous vibration of the PMG can lead to the levitation force attenuation. Table 2 collects the attenuation values of each case. The attenuation is the difference between the results obtained at the beginning (120 s) and the end (140 s) of the dynamic condition process, which can be seen in Figure 3b,d. We can find in Table 2 that with the increase of the frequency, the attenuation gets more obvious. This is because the flux inside the HTS bulk under higher vibration frequency is more intense. Other studies in the literature [15,16] have concluded that the levitation force is a little higher when not considering the thermal effect than when accounting for the thermal effect, because the attenuation caused by thermal loss is not considered. From this point of view, under the experimental vibration condition, the force attenuation without the thermal effect is a little smaller than the case considering the thermal effect, in which J_c retains some mathematic relationships with temperature T during the calculations. As shown in Table 2, the calculated force attenuation without considering the thermal effect is 0.009 at 2 Hz, and 0.065 at 60 Hz. However, the calculation results considering the thermal effect, such as 0.006 under $\alpha = 1$ at 2 Hz and 0.062 under $\alpha = 3/2$ at 60 Hz, are smaller. In addition, 0.010 under the linear function and 0.009 under $\alpha = 3/2$ at 2 Hz are almost the same, with 0.009 at 2 Hz without any thermal effect. Therefore, only the results calculated by Equation (7) with $\alpha = 2$ are satisfied and reasonable, which is the same conclusion by the quasi-static comparison in Figure 2.

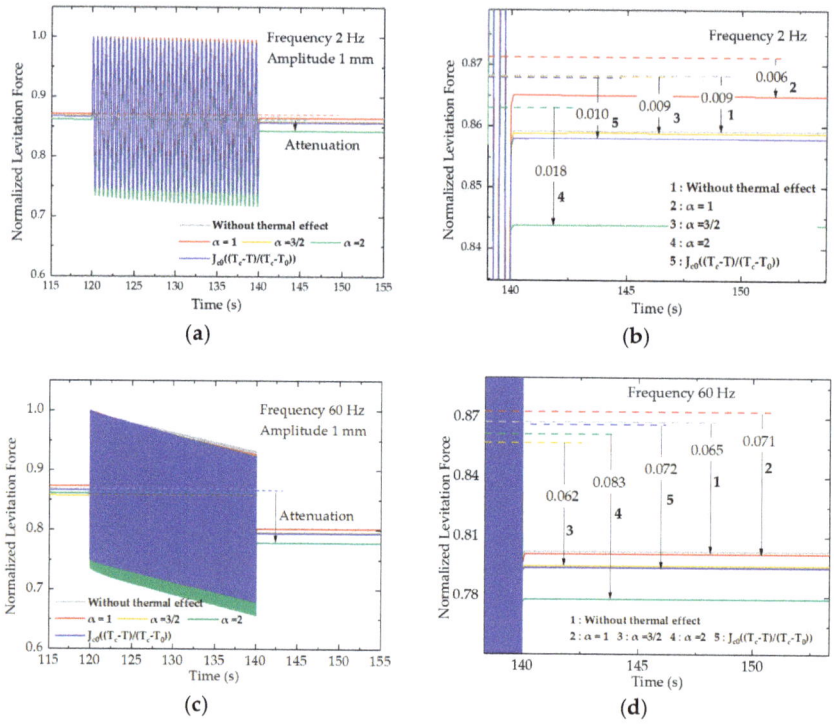

Figure 3. Normalized levitation force profiles by different J_c–T relationships of $J_c = J_{c0}$; $J_c = J_{c1}(1-(T/T_c)^2)^\alpha$, $\alpha = 1, 3/2, 2$; $J_c = J_{c0}((T_c-T)/(T_c-T_0))$ under (**a**) 2 Hz and (**c**) 60 Hz; (**b**) and (**d**) emphasize displaying the force attenuation caused by the vibration process under 2 Hz and 60 Hz, respectively. The dotted lines represent the force value at the end of Relaxation I (120 s). FCH = 30 mm, WH = 15 mm. 1, 2, 3, 4, and 5 represent five J_c–T calculation conditions.

Table 2. The normalized levitation force attenuation under different J_c–T relationships.

Frequency	Converted Speed	$J_{c0}((T_c-T)/(T_c-T_0))$	$J_c = J_{c1}(1-(T/T_c)^2)^\alpha$			Without Thermal Effect
			$\alpha = 1$	$\alpha = 3/2$	$\alpha = 2$	
f = 2 Hz	34 km/h	0.010	0.006	0.009	0.018	0.009
f = 60 Hz	1018 km/h	0.072	0.071	0.062	0.083	0.065

The current density and the temperature distribution of the HTS bulk under 60 Hz at the working height (15 s), the end of the first relaxation process (120 s), and the time when the PMG first reached the vibration peak (120.005 s) are shown in Figure 4a. It is seen that all the currents and the temperature rise appear from the bottom of the bulk, and the temperature gradually spreads to the entire bulk. All the modeling results indicate that the maximum temperature occurs at 15 s, when the PMG arrives at the working height after the quasi-static process. While the thermal effect by that tiny-amplitude vibration is small, it is concluded that compared to the small-amplitude high-frequency vibration of the

PMG, a long-distance movement between the FCH and the WH is more likely to lead to temperature rise, because it causes a distinct magnetic field variation despite the speed being only 1 mm/s. Figure 4b displays the maximum temperature inside the HTS bulk calculated by different J_c–T relationships under 60 Hz. It is seen that temperature rise occurs mainly around the time when the PMG first reached the working height, and the first upward moved 1 mm during the dynamic process. The inset in Figure 4b further indicates the temperature variation difference after the beginning of the vibration. Although the vibration process excites the second temperature rise, the temperature rise is much smaller than that in the first movement from the 30-mm FCH to the 15-mm WH, because the vibration amplitude is only 1 mm. Furthermore, the temperature curves kept decreasing gradually after the PMG first reached the peak position during dynamic conditions.

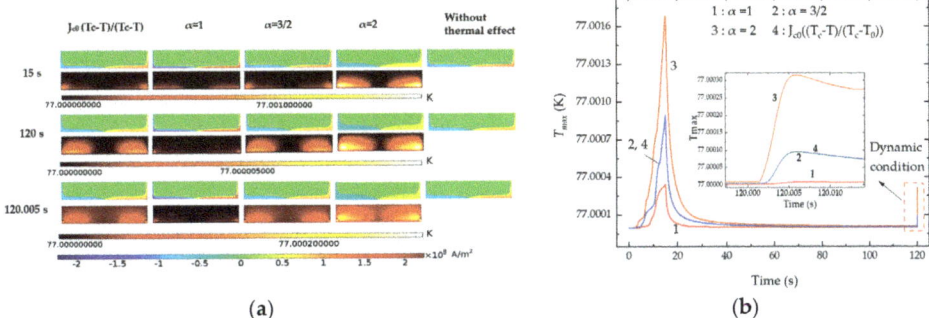

Figure 4. Comparison of different J_c–T relationships of $J_c = J_{c1}(1-(T/T_c)^2)^\alpha$ with $\alpha = 1, 3/2, 2$ and $J_c = J_{c0}((T_c-T)/(T_c-T_0))$ under 60 Hz. (**a**) Current density and temperature profile inside the HTS bulk at 15 s, 120 s, and 120.005 s; (**b**) maximum temperature T_{max} inside the HTS bulk versus time.

4. Conclusions

In this study, based on the verification of the model by the measured levitation force, we calculated the levitation force, the temperature distribution, and the current density distribution of the HTS bulk under dynamic conditions. Four functions describing the temperature dependence of the critical current density (J_c–T) were used in the calculations. The results were compared with the simulation case, without accounting for the thermal effect as well as the measurements. Considering the computation efficiency, we calculate 20 s for the dynamic operation, since the present computational condition is not capable of guaranteeing the long-time vibration calculation. The following conclusions are obtained:

1. The J_c–T function of J_c proportional to $J_{c1}(1-(T/T_c)^2)^2$ is more appropriate to reproduce the electromagnetic-thermo-force coupling characteristics of the HTS Maglev system.
2. According to the calculated dynamic levitation force, it is predicted that the simulated dynamic levitation force running at a high speed such as 1018 km/h would decrease 8.3% at the beginning, if the applied PMG field could be inhomogeneous along the running direction, similar to the particularly designed inhomogeneous PMG in the SCML-03 system. On the other hand, it is another significant research issue to design or optimize the working magnetic field homogeneity of the PMG for the high-speed HTS Maglev application.

Author Contributions: Data Curation, Y.H., J.Z.; writing and original draft preparation, Y.H., J.Z.; Experiment, Y.H., H.L.; Project Administration, Manuscript Review and Editing, J.Z.

Funding: This research was funded in part by the Sichuan Science and Technology Program (2019YJ0229 and 19ZDYF0436), and the State Key Laboratory of Traction Power at Southwest Jiaotong University (2018TPL_T06).

Conflicts of Interest: The authors declare no conflict of interest.

References

1. Wang, J.S.; Wang, S.Y.; Zeng, Y.W.; Huang, H.Y.; Luo, F.; Xu, Z.P.; Tang, Q.X.; Lin, G.B.; Zhang, C.F.; Ren, Z.Y.; et al. The first man-loading high temperature superconducting Maglev test vehicle in the world. *Phys. C* **2002**, *378*, 809–814. [CrossRef]
2. Okano, M.; Iwamoto, T.; Furuse, M. Running performance of a pinning-type superconducting magnetic levitation guide. *J. Phys. Conf. Ser.* **2006**, *43*, 999–1002. [CrossRef]
3. Sotelo, G.G.; De Oliveira, R.A.H.; Costa, F.S.; Dias, D.H.N.; De Andrade, R.; Stephan, R.M. A Full scale superconducting magnetic levitation (MagLev) vehicle operational line. *IEEE Trans. Appl. Supercond.* **2015**, *25*, 3601005. [CrossRef]
4. Deng, Z.G.; Zhang, W.H.; Zheng, J.; Ren, Y.; Jiang, D.H.; Zheng, X.X.; Zhang, J.H.; Gao, P.F.; Lin, Q.X.; Song, B.; et al. A high-temperature superconducting maglev ring test line developed in Chengdu, China. *IEEE Trans. Appl. Supercond.* **2016**, *26*, 3602408. [CrossRef]
5. Sotelo, G.G.; Andrade, R.D.; Ferreira, A.C. Magnetic bearing sets for a flywheel system. *IEEE Trans. Appl. Supercond.* **2007**, *17*, 2150–2153. [CrossRef]
6. Werfel, F.N.; Floegeldelor, U.; Rothfeld, R.; Riedel, T.; Goebel, B.; Wippich, D.; Schirrmeister, P. Superconductor bearings, flywheels and transportation. *Supercond. Sci. Technol.* **2012**, *25*, 014007. [CrossRef]
7. Matsushita, T.; Otabe, E.S.; Fukunaga, T.; Kuga, K.; Yamafuji, K.; Kimura, K.; Hashimoto, M. Weak link property in superconducting Y-Ba-Cu-O prepared by QMG process. *IEEE Trans. Appl. Supercond.* **1993**, *3*, 1045–1048. [CrossRef]
8. Yamasaki, H.; Endo, K.; Kosaka, S.; Umeda, M.; Misawa, S.; Yoshida, S.; Kajimura, K. Magnetic-field angle dependence of the critical current density in high quality $Bi_2Sr_2Ca_2Cu_3O_x$ thin films. *IEEE Trans. Appl. Supercond.* **1993**, *3*, 1536–1539. [CrossRef]
9. Wesche, R. Temperature dependence of critical currents in superconducting Bi-2212/Ag wires. *Phys. C* **1995**, *246*, 186–194. [CrossRef]
10. Matsushita, T. Flux Pinning in Superconductors. In *Springer Series in Solid-State Sciences*; Springer: Berlin/Heidelberg, Germany, 2007.
11. Braeck, S.; Shantsev, D.V.; Johansen, T.H.; Galperin, Y.M. Superconducting trapped field magnets temperature and field distributions during pulsed field activation. *J. Appl. Phys.* **2002**, *92*, 6235. [CrossRef]
12. Tsuchimoto, M.; Kamijo, H. Maximum trapped field of a ring bulk superconductor by low pulsed field magnetization. *Phys. C* **2007**, *464–465*, 1352–1355. [CrossRef]
13. Tsukamoto, O.; Yamagishi, K.; Ogawa, J. Mechanism of decay of trapped magnetic field in HTS bulk caused by application of AC magnetic field. *J. Mater. Process. Technol.* **2005**, *392*, 659–663. [CrossRef]
14. Tixador, P.; David, G.; Chevalier, T.; Meunier, G.; Bergera, K. Thermal-electromagnetic modeling of superconductors. *Cryogenics* **2007**, *47*, 539–545. [CrossRef]
15. Ye, C.Q.; Ma, G.T.; Yang, W.J.; Yang, Z.Y. Numerical studies on the dynamic responses of levitated high-temperature superconductor with a strongly coupled thermo-electromagnetic model. *J. Phys. Conf. Ser.* **2018**, *1054*, 012087. [CrossRef]
16. Huang, C.G.; Xu, B.; Zhou, Y.H. Dynamic simulations of actual superconducting maglev systems considering thermal and rotational effects. *Supercond. Sci. Technol.* **2019**, *32*, 045002. [CrossRef]
17. Zheng, J.; Huang, H.; Zhang, S.; Deng, Z.G. A general method to simulate the electromagnetic characteristics of HTS maglev systems by finite element software. *IEEE Trans. Appl. Supercond.* **2018**, *28*, 3600808. [CrossRef]
18. Huang, H.; Zheng, J.; Liao, H.P.; Hong, Y.; Li, H.; Deng, Z.G. Effect laws of different factors on levitation characteristics of high-T_c superconducting maglev system with numerical solutions. *J. Supercond. Nov. Magn.* **2019**, *6*, 1–8. [CrossRef]
19. Liu, L.; Wang, J.S.; Deng, Z.G.; Wang, S.Y.; Zheng, J.; Li, J. Dynamic simulation of high temperature superconductors above a spinning circular permanent magnetic guideway. *J. Supercond. Nov. Magn.* **2010**, *23*, 597–599. [CrossRef]
20. Liao, H.P.; Zheng, J.; Huang, H.; Deng, Z.G. Simulation and experiment research on the dynamic levitation force of bulk superconductors under a varying external magnetic field. *IEEE Trans. Appl. Supercond.* **2019**, *29*, 1–5. [CrossRef]

21. Liu, L.; Wang, J.S.; Wang, S.Y.; Li, J.; Zheng, J.; Ma, G.T.; Yen, F. Levitation force transition of high-T_c superconducting bulks within a maglev vehicle system under different dynamic operation. *IEEE Trans. Appl. Supercond.* **2011**, *21*, 1547–1550. [CrossRef]
22. Hong, Z.; Campbell, A.; Coombs, T. Computer modeling of magnetisation in high temperature bulk superconductors. *IEEE Trans. Appl. Supercond.* **2007**, *17*, 3761–3764. [CrossRef]
23. Paul, W.; Hu, D.; Baumann, T. Voltage-current characteristic between 10^{-13} V/cm and 10^{-3} V/cm of BSCCO and time decay of the magnetization. *Phys. C* **1991**, *185*, 2373–2374. [CrossRef]

© 2019 by the authors. Licensee MDPI, Basel, Switzerland. This article is an open access article distributed under the terms and conditions of the Creative Commons Attribution (CC BY) license (http://creativecommons.org/licenses/by/4.0/).

MDPI
St. Alban-Anlage 66
4052 Basel
Switzerland
Tel. +41 61 683 77 34
Fax +41 61 302 89 18
www.mdpi.com

Materials Editorial Office
E-mail: materials@mdpi.com
www.mdpi.com/journal/materials

www.ingramcontent.com/pod-product-compliance
Lightning Source LLC
LaVergne TN
LVHW070558100526
838202LV00012B/502